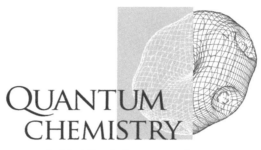

QUANTUM CHEMISTRY

A Unified Approach

2nd Edition

MERMAID
LEGENDS

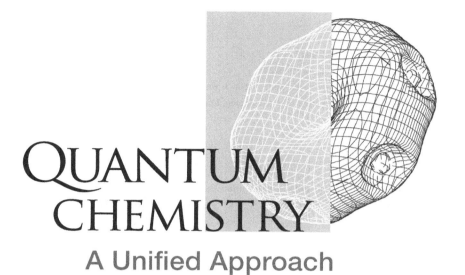

QUANTUM CHEMISTRY

A Unified Approach

2nd Edition

David B Cook

University of Sheffield, UK

Imperial College Press

Published by

Imperial College Press
57 Shelton Street
Covent Garden
London WC2H 9HE

Distributed by

World Scientific Publishing Co. Pte. Ltd.
5 Toh Tuck Link, Singapore 596224
USA office: 27 Warren Street, Suite 401-402, Hackensack, NJ 07601
UK office: 57 Shelton Street, Covent Garden, London WC2H 9HE

British Library Cataloguing-in-Publication Data
A catalogue record for this book is available from the British Library.

QUANTUM CHEMISTRY
A Unified Approach
(2nd Edition)

Copyright © 2012 by Imperial College Press

All rights reserved. This book, or parts thereof, may not be reproduced in any form or by any means, electronic or mechanical, including photocopying, recording or any information storage and retrieval system now known or to be invented, without written permission from the Publisher.

For photocopying of material in this volume, please pay a copying fee through the Copyright Clearance Center, Inc., 222 Rosewood Drive, Danvers, MA 01923, USA. In this case permission to photocopy is not required from the publisher.

ISBN-13 978-1-84816-746-9
ISBN-10 1-84816-746-6

Typeset by Stallion Press
Email: enquiries@stallionpress.com

Printed in Singapore.

Contents

Preface to the Second Edition

The main text of the first edition has been changed very little. I have improved some poor diagrams and added one or two new ones. This edition has had several typographical and other errors corrected thanks to Tomas Sordo and (as usual) Patrick Fowler. Grateful thanks to both these friends for a careful reading of the text. Since new material has been added and I have made minor textual corrections to the material in the first edition, I must apologise in advance for any remaining errors, which are entirely my responsibility.

The additional material is of two types. I have added several appendices containing material which may be omitted but which does deal with several important points. These appendices have been added partially for amusement (Appendix F), partially in self-defence (Appendices E and H) and partially to give some support to concepts which, if included in the main text would have hindered the main thrust of the presentation (Appendices G and I). If there is any unifying theme in the new appendices it is the result of my thinking about the sort of persistent oversights and errors which have been tacitly ignored by many writers and which have become part of the conventional wisdom. The rôle of electron spin in valence theory, the general lack of emphasis on inter-electron repulsion and the relationship between these two effects are consequently prominent in these additions.

Each chapter has an Assignment Section containing 'problems' which might be usefully attempted to improve the understanding of the new material in that chapter. This book was always intended to be a *qualitative* presentation of some material in quantum chemistry, and this has been emphasised by the type of problem in the assignments. They are, in the main, points to think about and discuss. The best use of these assignments is in a classic 'University Tutorial' environment; the students prepare some written thoughts and these are discussed, preferably in a peer group with a tutor or tutors. My overall experience, both as a tutee and as a tutor, has convinced me that setting a number of short (often numerical) problems to be marked and returned often has the rather unfortunate result that the students can usually do all the problems but still do not have a firm grasp of the ideas the problems are supposed to illustrate. I have provided some guidelines to the sort of points which ought to come out of this work at the end of the book. Not all the problems are specifically chemical and the answers are not always to be found in the text. There are one or two additions to the text among these points which might be used to deepen the understanding of interested students.

David B. Cook, Sheffield, January 2011

Preface

One of the earliest books about the qualitative theory of the electronic structure of molecules — *Valence Theory* by John Murrell, Sid Kettle and John Tedder (1965) — was a collaborative project by a physical chemist, an inorganic chemist and an organic chemist, all from the Department of Chemistry at the University of Sheffield. The very idea of scientists with such disparate backgrounds co-operating on such a venture seems scarcely credible today. I have been teaching quantum chemistry for forty years at all levels; from introductory courses for medical students to graduate lectures for Ph.D. chemists and physicists. During that time there has been a gradual movement away from the idea that there can be a qualitative theory which can be accepted and used by all branches of chemistry (or even by a wider consensus). An amusing example is provided by the concept of a dative bond: for contemporary organic chemists such a thing does not exist, yet much of modern inorganic chemistry is dominated by the idea.

Organic chemists have, over the years, either abandoned the idea of a quantum-based theory or have used the nomenclature of quantum theory to develop a series of rules which no longer refer to the physical laws that underlie chemical phenomena.

Inorganic chemists have, in general, been hampered by their history. The idea that the bonding in transition-metal molecules is ionic and only needs 'correcting' for quantum-mechanical effects, together with the use of symmetry-adapted molecular orbitals (MOs) for this correction have prevented the growth of a qualitative theory of the wealth of novel electronic structures to be found in inorganic molecules.

Physical chemists, who are in the best position to provide common ground, have unfortunately remained aloof from these difficulties, concentrating instead on quantitative methods and, in extreme cases, insisting that there is no such thing as a chemical bond, only calculated electron distributions!

In this work I have tried to explain things in terms of the interactions amongst charged particles; something which, if it *has* been attempted before, has not been tried for many decades. The first two or three chapters are devoted to providing the basis for the simplest and most commonly-used prescriptions for an orbital *description* of the electronic structure of molecules. Of course, explanations can only go back so far and have to stop somewhere. What I have attempted is to present a theory of valence which is based on the idea that the factors which shape molecules and their interactions are based on two main sources: experimental observations and the laws of nature — electrostatic interactions between particles and Schrödinger's mechanics — and not rules of thumb involving atomic orbitals. Some of the things I have said are controversial and others are stated without proof because of the nature of this work. A more thorough case for the relationship between probability and the interpretation of orbitals is to be found in my *Probability and Schrödinger's Mechanics* (World Scientific, 2002), and the technical support for the computational methods used are given in *Handbook of Computational Quantum Chemistry* (Dover, 2005).

David B. Cook, Sheffield, 2007

Acknowledgments[1]

Producing this book has involved a great deal of computing and a lot of hacking.[2] The vast majority of the basic quantum-chemistry codes have been taken from that utopian enterprise, the Quantum Chemistry Program Exchange (QCPE), set up by Harrison Shull in 1962 and overseen by Richard Counts for many years. Like so many other things, the idea of a free interchange of software between scientists has been overtaken by 'market forces' and 'entrepreneurial talent' and, as I write this, QCPE is in its last months; closing through lack of funding. I have changed and developed these programs over the years so as to keep up with the changes in the Fortran programming language and with my own scientific requirements. While no longer useful for the large-scale production work that is the bread and butter of contemporary computational chemistry, they are still of value in investigating the details of molecular electronic structure, which has been my concern here. Grateful thanks, therefore, to those pioneers of the open source software movement.

The preparation of this work would not have been possible without those pillars of the computing world: Linux and the teTeX distribution of LATEX. Naturally, the production of a scientific work is inconceivable without emacs and all its support. These works are a real source of civilisation in the world.

My friend and colleague Patrick Fowler read some drafts of parts of the work and gave me his usual valuable and forthright opinions, and we have discussed some of the more controversial points at some length; I thank

---

[1]This was written for the first edition in 2007.

[2]The word hacking was used in the first edition when it was used in the computing world to mean nothing more sinister than rewriting parts of a program.

him for the generous way in which he has always been willing to look at my work.

The editors of Imperial College Press have been tolerant — above and beyond the call of duty — of my prevarication and sloth for the past several years while the manuscript was in preparation, and I thank them for not giving up on me.

Finally, I must thank my wife Irene for her constant encouragement, threats and ultimatums; without her nothing would be achieved.

Chapter 1

How Science Deals with Complex Problems

The theory of the electronic structure of molecules is like many problems in science; the basic laws underlying the problem are very well known, but the way in which these laws are to be applied to yield results which are meaningful in a particular context is not obvious. This situation is, perhaps, most familiar in the application of Newton's mechanics to the motions of the solar system. Just as it is not possible to predict the number of satellites which actually orbit Jupiter, we cannot obtain a description of the electronic structures of molecules solely from the basic laws; we must supply additional ideas from a knowledge of the chemistry involved. The main idea used here is the breakdown of the overall system into a series of interacting sub-systems.

Contents

1

1.1   Introduction: Levels in Science

It is commonly said that Newton's laws give an essentially perfect description of the internal motions of the solar system; the motions of the moons around the planets and of the planets around the sun. There is similar confidence among some physicists that Schrödinger's mechanics (quantum theory) is capable of providing a complete quantitative account of chemical phenomena. Both of these opinions are wrong. The ways in which they are wrong are as important to understand as the sense in which they are thought to be right.

Let's see what would be involved in calculating the orbits of the various bodies in the solar system. We would need:

- Newton's laws of motion (principally $F = ma$).
- The number of bodies and their masses (sun, planets, moons, main asteroids — ignore small bodies and dust etc.).
- The law of interaction between the bodies (gravity, just depending on the masses of the bodies and their distances apart).

We have all this information so we can now set to and calculate the paths of all the bodies in the solar system.

Compare this with the information needed to calculate the distributions and energies of the electrons in (say) the ethanol molecule:

- Schrödinger's laws of motion (the Schrödinger equation).
- The number of bodies, their masses and charges (nuclei and electrons).
- The law of interaction between bodies (electrostatics, Coulomb's law of attraction and repulsion).

Again, all this information is known so we can go ahead and calculate everything about the ethanol molecule.

The two cases are remarkably similar: a few dozen bodies, a known law of interaction between them and the correct laws of motion. But in neither case can we actually go ahead and get the desired result because we do not know how the systems of bodies are *organised*. In order to even begin we need to know lots of 'large-scale' information; for example:

- How many moons each planet has.
- The connectivity of the ethanol molecule (which atoms are 'bonded').

What is common to both cases is the fact that each of these relatively simple mechanical systems is, in fact, composed of a number of smaller sub-systems, and it is precisely the way in which the overall systems are broken down into smaller sub-systems which must be known before a detailed account of the motion can be even attempted. For example, I am sure that there will be thousands of possible solutions of Newton's equations where the planets have no moons and all the bodies orbit the sun as planets; conversely there will be solutions of the dynamical equations where Venus has a moon and Mars has none, and many, many more in addition to the actual, real case.

However, what *is* true is that (in both the solar system and the molecule), *when the breakdown of the whole system is known* and the correct mechanics is applied to the sub-systems, the motions of the sub-systems and hence of the whole system can be calculated with great accuracy. In other words, when studying any system of even moderate complexity, a great deal of information is required at what one might call a 'high' level before one can even think about applying the basic, 'low-level' laws which are commonly said to be able to explain the structure and motions of the system.

The upshot of these considerations is that, in any science which is not concerned with the most basic phenomena, there is an all-important analysis to be made before any quantitative calculations can be attempted. The system under study must be analysed into the essential sub-systems of which it is composed. Now here's the problem: how do we decide which are these 'essential' sub-systems into which the whole system may be divided in the most useful and revealing way? In the case of the solar system the analysis is easy, partly because we can literally see what the most useful sub-systems are: we simply take a telescope and see that (for example) Jupiter *and all its moons* orbit the sun together *as a group* while simultaneously undergoing their own 'local' internal motions (the moons orbiting the planet). The situation is similar for all the other planets; they are either orbiting alone or taking their moon(s) with them. So, we might take the following strategy:

- Use the mass of each planet together with its moons and use Newton's equations to calculate the orbit of this whole group around the sun, ignoring the other planetary groups.
- Then calculate the orbits of the moons around each planet, ignoring the sun and the other planetary groups.

- Finally calculate the mutual interactions amongst the planetary groups — how Jupiter influences the motion of Saturn and its moons etc.

The whole idea here is to divide up the system into sub-systems which are mutually interacting, but in a way that gives a good deal of *independence* to each sub-system so that they are recognisable as sub-systems whose motions are not completely disrupted by the presence of the other sub-systems; the orbit of the moon around the Earth *is* affected by Jupiter, but not by much.

> This analysis into substructures has to be carried out even if no actual detailed calculations are to be done; breaking down a complex structure into its component parts is the way we *actually think* about these structures. No-one pictures the solar system in their minds as just a set of dozens of bodies, nor do they picture a molecule as just a collection of electrons and nuclei.

This is all, of course, very obvious because of the clear distinction between the planetary groups and the enormous distances between them. But what about the problem of the motions of electrons and nuclei in a molecule? All this complicated motion occurs in a very small region of space, and all the bodies are *charged* and the interactions amongst charged particles are billions of times stronger than gravitational interactions.

> We have to decide what are the optimum subsystems into which a molecule is to be divided. In other words, 'What are molecules made of?'

In this work I shall be concerned with *qualitative* descriptions of the electronic structure of molecules, and only using detailed calculations 'in the background' to provide illustrations; but the large-scale analysis still has to be done to make sense of the structure of even quite small molecules.

1.2   What Are Molecules Made of?

Anyone with even a smattering of chemical experience can provide a list of possible answers to the question 'what are molecules made of?', and these answers might be sorted 'by size':

(1) Molecules are composed of functional groups and structural backbones;
(2) Molecules are composed of atoms;

(3) Molecules are composed of bonds, lone pairs and delocalised electron structures;

(4) Molecules are composed of electrons and nuclei;

and all these answers are correct. Each of these answers is correct *for a particular way of thinking about molecules*. The question which we must address is:

> What are the substructures of a molecule which are optimum
> for a meaningful description of their electronic structure?

In this context the first of the above answers can be discarded, since it is the way that chemists concerned with the practical problems of synthesis might break down the structure of a molecule in their minds. The last of the answers looks more like a description of the structure of a metal than that of a molecule; it is more a physicist's analysis than a chemist's and will also be discarded.

There are now two serious contenders for the method to use: are molecules made of atoms or are they made of bonds, lone pairs and other electronic structures? We might pause to think about what we mean by 'made of'; we might mean 'are constructed from', or 'can be made to emit', or 'can be shown to contain'. What is certainly true is that molecules can be constructed from atoms,[1] and molecules may be broken up into atoms,[2] whereas we cannot isolate (e.g.) a lone pair or a π-bond from a molecule, and we certainly cannot use such objects to make molecules. Colloquially we might say:

> Are molecules made of the subsystems like walls are made of
> bricks, or like omelettes are made of eggs?

In fact, the real question is a little more precise than the one I have asked; what we really require is the answers to the questions:

> What substructures are the *electronic structures* of molecules
> made of? What is the best choice of electronic substructure
> in order to make sense of the structure and properties of
> molecules?

When posed in this way the choice seems easier but, as I shall try to show, the choice is never obvious, and we shall always have to have answers both 2

---

[1] With a great deal of difficulty, though.

[2] Again, not easily.

and 3 in mind all the time. In fact, it will become clear that it is absolutely essential to keep these two points of view and the tension between them in mind all the time if we want to create a realistic qualitative theory of the electronic structure of molecules; a theory that can be made quantitative enough to be sure that we are on the right lines.

1.3   Interactions Between Atoms

If we take the point of view that molecules are made of atoms then we can think about what might happen when atoms approach one another; what are the processes which occur[3] under these circumstances? First of all we need to think about what atoms actually *are* at the simplest possible level; let's say for the moment that an atom is just a (relatively) massive positively-charged nucleus surrounded by a cloud of (relatively) light negatively-charged electrons, with the atom being overall electrically neutral (the charge on the nucleus — the atomic number — being the same as the number of surrounding electrons). What can happen to such structures when they are brought close enough together for them to be influenced by each other's distribution of electrical charges?

First of all we can treat one atom as stationary in the sense that we can think of the nucleus of one of the interacting atoms as being at a fixed point in space, surrounded by its electrons as the other atom approaches. The electrons of this atom are, of course, in constant motion around the nucleus and are kept captive by the attraction between the (positively-charged) nucleus and their own negative charge. How will the approach of another, similar, system of charges affect the atom? There are, at least, two effects which come to mind immediately:

(1) The target atom's electron distribution will be *polarised* by the approach of any charged body, or rather the *collection* of charged bodies, which is the approaching atom. The motions of the atom's electrons will be changed by the combination of attractions and repulsions between the charged particles in target atom and the approaching atom. Or to put it in more convenient language, the *distribution* of the target atom's electrons will be distorted. That is, the electron

---

[3]Or we might think of as occurring.

distribution around the atom will no longer be spherically symmetrical in the presence of another atom.[4]

(2) Similarly, due to the electrostatic interactions between the two sets of particles of which the atoms are composed, the distribution of each atom's electrons may be changed in another way: the overall '*size*' of the electron distributions may change due to some of the electrons moving closer to or further from their own nucleus as they 'feel' the attractions and repulsions from the other atom. Each atom may get slightly bigger or smaller due to the presence of the other atom.

Naturally, these two effects do not happen *separately*; what must happen in practice is that the electron distributions interact with each other and distort with any expansion or contraction occurring *simultaneously* with any polarisation of the original distribution. But we can separate these two effects in our minds to try and understand each one individually.

Finally, as the two atoms approach closely, some of the electrons of both may be sufficiently attracted to both nuclei *simultaneously* to disrupt their original motions around their 'own' nucleus completely and take up an altogether new set of motions ('distribution') around *both* nuclei. When this happens, the two atoms are involved in a completely new set of circumstances; they are held together by these electrons which are attracted strongly to *both* nuclei — in a word they are *shared* between the two nuclei — and they hold the two nuclei together in a combined system which we recognise as a chemical bond.

Returning to the two answers to the question 'what are molecules made of', we can now see the way in which answers 2 and 3 on pages 4 and 5 are intertwined; we can easily say that our simple considerations could be interpreted in two ways:

(1) The two *atoms* are joined by a chemical bond or;
(2) The electron distribution around the two *nuclei* has taken up a new structure; some of the electrons are distributed in a similar way around the nuclei but some electrons have changed their distribution radically.

Both descriptions are correct; the first is just a useful *name* for what has happened and the second is a mechanistic *explanation* of what *caused* the phenomenon. In this work we shall be concerned mostly with the second

---

[4]Of course, the target atom will distort the distribution of the electrons on the approaching atom in a similar way.

of these interpretations and, sooner or later, we shall have to decide what we mean by the term 'explanation'.

1.4   The Simplest Examples: H_2 and LiH

1.4.1   *The hydrogen molecule*

Before starting to get a handle on a quantum-mechanical theory of these processes, let's look at the simplest possible case: the hydrogen molecule. This is simply a look-ahead at some results which will be explained later, but which enables us to see if we are even approximately on the right track. No attempt will be made yet to say how these results were obtained, they must be taken on trust for the moment.

The way that quantum-mechanical calculations are actually performed has some interesting advantages which we shall meet in due course, but for the moment we simply note that it is actually possible using standard methods of quantum chemistry to perform calculations which approximately separate the two effects discussed in answers 2 and 3 on pages 4 and 5, and see the effect of the mutual polarisations and expansions/contractions of the two atomic electron distributions separately.[5] The table below gives the bond energy of the H_2 molecule as calculated allowing the two effects separately and together. For interest the experimental bond energy is given.

Type of Calculation	Calculated Bond Energy (kJ mol^{-1})
Unchanged Atomic Orbitals	190.86
'Size-Optimised Atomic Orbitals'	310.33 (change is 119.47)
'Polarised-only Atomic Orbitals'	229.53 (change is 38.67)
'Size-Optimised and Polarised AOs'	326.36 (change is 135.5)
Experimental Bond Energy	435.56

It is quite easy to see in this particular case that the effect of changes of size (contraction, in this case) of the electron distributions of the two atoms on the calculated bond energy is noticeably larger than the effect of their mutual polarisations. This fact is in line with the fact that the energy required to polarise the hydrogen atom is quite high. Obviously, the more polarisable the atom, the more it *will* be polarised in a molecular

---

[5]For the experts, this is done by choice of orbital basis.

environment. It is also easy to see that the calculations do not satisfactorily reproduce the numerical value of the bond energy — more on this later.[6]

As the contour[7] diagrams below illustrate, the *polarisation* of each H atom is a very slight shift of the contours in the direction of the other atom. In fact, the change in size (*contraction*) of each atom's electron distribution is too small to be seen on the scale of these contours; it is a very slight 'pulling in' of the contours when shown on a much-enlarged scale. In the diagram the positions of the two hydrogen nuclei are given by the small crosses, so there are two diagrams of the *whole molecule* given below, showing the fact that the electron distribution of *each atom* overlaps both nuclei. So, the original electron distribution of the left-hand hydrogen atom is shown in the left-hand diagram together with the positions of both nuclei, and the contracted, polarised distribution is shown on the right-hand atom of the right-hand diagram. The two almost identical contours on the right-hand diagram are the contracted distribution and the contracted, polarised distribution. In the case of the electron in the hydrogen atom, which is very tightly bound to its nucleus, there is nothing much to be seen.

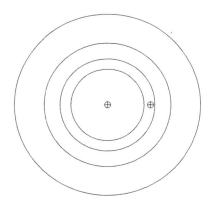

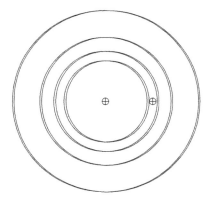

Atomic electron distribution
on left-hand atom

Contracted, polarised distribution
on left-hand atom

[6] Notice that the two effects (as calculated here) are not quite independent; the sum of the two 'independent' changes is larger than the combined total change; the method does not succeed in completely separating the two effects, something else to look at later.

[7] See Appendix A at the end of this chapter for hints on how to interpret contour diagrams.

Finally, contour diagrams of the electron distribution when the two contracted, polarised atomic distributions (left-hand diagram) are allowed to interact fully to form the molecule H_2 (on the right). Notice how the full interaction — when both electrons are attracted to both nuclei — is more compact than the original atomic distributions.

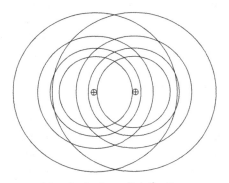
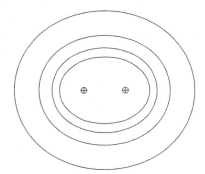

Atomic electron distributions
of the two atoms superposed

Electronic distribution of
the molecule

Overall, the results of these calculations are hardly definitive; the effects we predicted are present but scarcely spectacular.

1.4.2   *The lithium hydride molecule*

The lithium hydride molecule is much more interesting in this context because it has two *different* atoms, the lithium atom having one of its electrons (the 'outer' one) which is much less tightly bound to its nucleus than the electron of the hydrogen atom and so is much easier to move around.[8] The effect of expansion or contraction of the atomic charge distributions is difficult to predict on the basis of simple ideas but, if the outer electron of the lithium atom is easier to remove from the atom or excite, it should also be easier to distort or polarise its distribution. Let's see what happens.

---

[8]The energy required to remove the outer electron of a lithium atom is only 40% of the energy needed to remove the hydrogen atom's electron, and the energy required to excite the lithium $2s$ electron is also much less than that required to excite the hydrogen $1s$ electron.

Below are very simple contour diagrams of the distribution of the outer
($2s$) electron of the lithium atom as it is approached by a hydrogen atom;
only the position of the hydrogen *nucleus* is shown (vertically above the
lithium). The hydrogen atom's electron distribution — which is not visibly
polarised on the scale of these diagrams — is omitted for clarity. The
internuclear distances (R, which are the distances between the lithium
nucleus and the small dot marking the hydrogen nucleus) are 6, 5, 4, 3, 2
and 1.64 (the experimental bond length) respectively.

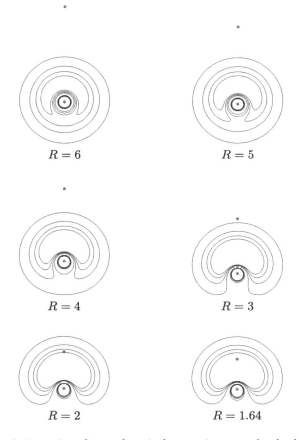

In this case it is quite clear what is happening: as the hydrogen atom
approaches, the polarisable outer electron distribution of the lithium atom

is attracted to the approaching atom. At very large distances the lithium atom senses nothing and remains spherical but, as the hydrogen atom comes within the effective electrostatic range of the other atom:

- The electron distribution is pulled towards that atom; the spherical appearance is progressively distorted, and
- The electron distribution is concentrated more in the inter-atomic region; more contours appear in that region.

This polarised electron distribution is attracted to the hydrogen nucleus — and repelled by the hydrogen's electron, of course; the net effect is attraction — and the (much less polarised) hydrogen atom electronic distribution also experiences net attraction to the lithium atom. At close enough approach, the overall effect is the same as in the hydrogen molecule:[9]

> When these two atoms approach each other closely enough, the electron distributions of the outer electrons of the two atoms are completely disrupted by their changed environment and take up a new distribution around *both* nuclei, which binds those nuclei together in a single chemical bond.

For completeness, the 'molecular' electron distributions at the distances in the above diagrams are given below:

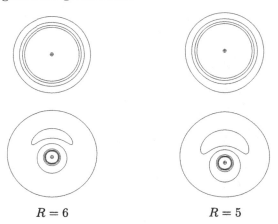

$R = 6$ $R = 5$

---

[9]The 'outer' electron of the hydrogen atom is its only one, of course!

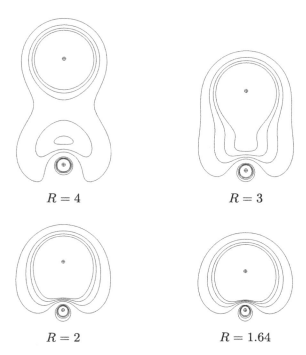

$R = 4$ $R = 3$

$R = 2$ $R = 1.64$

It is clear from these simple diagrams that there is little tendency for a molecule to form at all at the two largest distances, $R = 5$ and $R = 6$. All that happens is that the *atomic* electron distributions are distorted by each other's presence. Only when the two atoms approach to a distance of $R = 4$ or less do we see marked changes in the electron distribution, indicating something new happening. When the internuclear distance is reduced to around $R = 2$ or $R = 1.64$, the electron distribution is obviously that of a combined molecule; a single bond.

1.4.2.1 *What about the other Li electrons?*

In the above discussion we have said nothing at all about the 'inner' two electrons of the lithium atom. These two electrons (the $1s^2$ shell) are very tightly held by their close proximity to the lithium nucleus (charge $+3$). They are *extremely* difficult to disturb; their distributions are not affected by the presence of an approaching hydrogen atom. In fact, their distribution is not appreciably affected by molecule formation.

1.4.2.2 *What about the nuclear repulsions?*

What has been said about the way the pairs of electrons in the hydrogen and lithium hydride molecules become re-arranged when the atoms approach one another is all very well and, although it is nice to see our predictions on page 7 verified, we cannot yet tell whether or not these changes are sufficient to hold the two nuclei together. This is because only those changes which might be thought to hold the nuclei together have been considered. The calculations illustrated by the diagrams were done by simply *fixing* the nuclei at the quoted distances and seeing how the electrons responded. What about those effects, if any, forcing the atoms apart? In other words, what about the fact that the two *nuclei* repel each other because they are both positively charged?

Will the re-arrangement of the electrons described be sufficient to *overcome* this repulsive force? Unfortunately the answer is 'sometimes it is and sometimes it isn't', depending on the individual case. Simple evidence for this is easily found by going to the stores in any chemistry laboratory and asking for (cylinders of) hydrogen and helium. In the one case — hydrogen — the cylinder will contain molecules, and in the other — helium — one will find atoms. The stubborn refusal of helium atoms (and the other 'noble' gases: neon, argon etc.) to form molecules is just as much in need of explanation as the eagerness of hydrogen to bond with practically any atom. We shall have to look at this in detail later.

1.4.3 *Comments on H₂ and LiH*

How do our original comments on what might happen when atoms approach hold up in view of these results? Remember that we have not yet found out the theory and methods used in these calculations, that is for later. For the moment they are just presented as illustrations to be taken on trust.

In the case of the hydrogen molecule there is nothing much to see; two of the three distinct effects we thought might happen are hardly visible:

(1) The change in overall size of the atomic electron distributions is too small to see on the contour diagrams, although the calculated bond energies show that it is present. The detailed numerical results of the calculation show that this effect does occur; the atomic electron distribution does, in fact, contract on molecule formation.

(2) The polarisation of each atom's electron distribution by the other's
presence *is* visible but it is a very small effect in the case of hydrogen

(3) However, on very close approach, the two atoms' electron distributions
are disrupted and a new, molecular, electron distribution emerges corresponding to what we see experimentally; two hydrogen atoms are
more stable as a molecule than as separate atoms.

The lithium hydride molecule presents a completely different picture.
This time:

(1) The polarisation of the more loosely-bound outer electron's distribution is clearly visible in the diagrams, changing from the original atomic
spherically-symmetric distribution to a very directional distribution
with close approach of the hydrogen atom.

(2) The change in size of each atom is, however, again only to be detected
by the change in bond energy being too small to notice on the contour
diagrams.

(3) Just as in the case of H_2, the electron distribution of the two atoms
takes on a distinctly new, molecular form at small internuclear distances.

(4) There is an 'inner shell' of two $1s$ electrons on the lithium atom which
is basically unchanged on molecule formation; these electrons are held
so tightly by their 'parent' nucleus that their distribution is scarcely
affected by their new environment.

1.5 How to Proceed?

This very simple introductory chapter has raised many more questions
than it has answered, and has not touched on the major questions raised
by the experimental facts of chemistry:

- What about *polyatomic* molecules? How are atomic electron distributions changed and combined when an atom is bonded to more than one
other?

- Why does bonding *saturate*? Why does (e.g.) a carbon atom overwhelmingly from four bonds and not three or six?

- How is the *shape* of polyatomic molecules determined? The simple
diatomics we have looked at have no choice about their shape.

These kinds of question are all capable of being given quantitative answers by the methods of quantum chemistry and, more importantly from the point of view being taken here, the answers to these questions can all be understood *qualitatively* using the techniques which will be developed.

What I have been concerned with here has been the most *basic* of all the methods we shall have to use to understand the electronic structure of molecules; the question raised on page 5:

> Before we even to begin to think about the structure of a complex system like a molecule we have to use some experimental information to decide on the nature of the basic substructures within the molecule — what are molecules made of, what is the most fruitful breakdown of the overall structure into small structures? And, how do these substructures interact?

The examples of Section 1.4 provide some pointers:

- The hydrogen molecule is the simplest possible (chemically stable[10]) molecule and its electronic structure is such that *all* — i.e. both — its electrons have taken up a distribution around both nuclei to form a single chemical bond. There are *no* substructures in the hydrogen molecule; it can be seen as a *pair* of bonded atoms or as a *single* electron-pair bond according to the point of view.
- Lithium hydride is more interesting. We have seen that the electronic structure of the molecule consists of two substructures:

  (1) The inner shell of the lithium atom (the two electrons in the $1s^2$ shell) which has basically the same distribution in the atom and in the molecule.

  (2) The two electrons involved in the chemical bond, whose distribution (and energy) are completely different in the molecule from their separate atomic distributions.

So, to labour the point, the six-particle system, which is the lithium hydride molecule (two nuclei and four electrons), can be seen to consist of two two-electron substructures, which interact of course but whose basic structures are insensitive to changes in each others' detailed structure.

---

[10]The one-electron molecular-ion H_2^+ does exist but it is not the sort of thing that one can get in a bottle.

The last point is the so-called 'core-valence separation' and is the most obvious example of the more general point of view adopted here:

> The electronic structure of molecules is composed of 'environment-insensitive' substructures. As we shall come to see later, the ones of interest are often just *pairs* of electrons (bonds, lone pairs) but may be larger groups (inner shells, conjugated π systems).

If we return to the analogy with the solar system we can see that:

> Before we can calculate the details of the motions of the planets and their satellites, we have to know, among other things, how they are grouped together (how many moons has Jupiter?) and the fact that their orbits generally lie in a plane. Only then can we bring to bear the massive power of Newton's mechanics to predict their motions with enormous accuracy.

Likewise:

> Before we can compute the detailed distributions of the electrons we have to know (among other things) how the electrons are grouped, what relationships their distributions bear to the electron distributions in the separate atoms. Only then can we call on the big guns of Schrödinger's mechanics to calculate electron distributions and molecular shapes which can, in fact, be done with almost arbitrary accuracy.

So, the first thing to do is to see what we actually *do* know about the electronic structure of atoms and what we know about the ways in which atoms combine to form molecules. This means another chapter.

1.6 Assignment for Chapter 1

(1) Galileo is famous for his challenge to the authority of official scholars by appealing to experiment rather than the writings of classical philosophers. His most famous experiment was to show one feature of gravity by dropping two objects of very different mass from the Leaning Tower of Pisa and have it verified that they both hit the ground

at the same time. The experimental 'law' that this demonstrates is that:

> Objects of different weights (mass) fall to earth in the same time, i.e. they are accelerated by gravity by the same amount.

This is still not immediately obvious since gravity obviously exerts a much greater *force* on the heavier body, and Newton's Law says: the more force, the more acceleration.

Now, here's the problem:

Suppose that you were a loyal member of the establishment (the church, say) and were opposed to the very *idea* that the ideas which had been passed down for centuries — all the way from Aristotle 2000 years earlier — and wished to cast doubt on Galileo's 'law'.

> (a) Without moving from the Cathedral square, how, just by looking around at the happenings there, could you provide some simple examples which *disproved* Galileo's claim?
>
> (b) What kind of objects would *you* use to drop from the tower to strengthen your disproof?

Why are Galileo's experiments and yours conflicting and how can the conflict be resolved? Use your examples and observations to provide a more exact formulation of 'Galileo's law'.

(2) Are you convinced of the idea that one way to answer the question 'what are molecules made of?' is to say that they are made of nuclei and electronic substructures? Or, would you prefer to always think of molecules as made of atoms? The first idea is more 'abstract' than the more familiar second answer and depends on things which cannot be 'seen' in the laboratory; is this OK?

Discuss this with your colleagues and your tutors to see if you can come to a common point of view, and if not, why not?

Appendix A

How to Interpret 3D Contours

*Quite a lot of use will be made of representations of elec-
tron distributions and orbitals throughout this work. It
requires a little effort to make sure that these diagrams
are properly understood.*

Contents

A.1   Thinking in 3D

We are familiar with representing a function of one variable by using a 2D
graph; for the function $f(x)$ (say) we define a new variable y by

$$y = f(x)$$

and, using perpendicular axes, we plot y upwards and x along; this defines
a *curve* in 2D space. In this way we can get intuitive pictures of the way
in which the function 'behaves' as well as interpretations of the derivative

$$\frac{dy}{dx} = \frac{df(x)}{dx}$$

as the slope of the graph, and of the definite integral

$$I = \int_a^b f(x)dx$$

as the area under the curve between $x = a$ and $x = b$. Here is a simple graph of a function of x ($y = x^3 + 1$) together with the tangent (slope at $x = 1$: $dy/dx = 3x^2 = 3$), and the integral from $x = 0.5$ to $x = 1.5$ is the area bounded by the two vertical lines, the x axis and the curve:

$$\int_{0.5}^{1.5} (x^3 + 1)dx = \left[\frac{x^4}{4} + x\right]_{0.5}^{1.5} = 2.25.$$

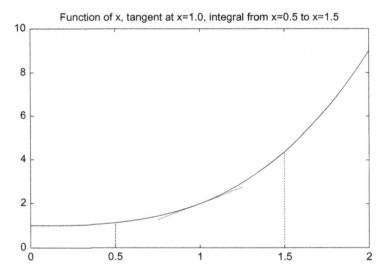

Function of x, tangent at x=1.0, integral from x=0.5 to x=1.5

It is less familiar but quite common to represent a function of two variables (x and y, say) using a 3D graph. Again, for the function $g(x, y)$ (say) we define a new variable z by

$$z = g(x, y)$$

and use two perpendicular axes in a horizontal plane to represent x and y and plot z 'upwards' perpendicular to both x and y. This defines a *surface* in ordinary 3D space.

Now, the problem is 'how to represent a function of three variables graphically'. From the examples above it is clear that, to represent a function of n variables ($n = 1$ or 2) one needs a space of $(n + 1)$ dimensions (2 or 3). But there are only *three* spatial dimensions in the world. We have to use experience from other disciplines. Geographers solved this problem long ago but in two dimensions; they are used to representing the topography of a country on a 2D map by using *contours* to represent the height of the land above sea level. They can represent a function of two variables — height above sea level as a function of latitude and longitude — on a *2D* surface; simply a relief map.

Opening any atlas at a page with a relief map of any country explains how this is achieved:

- Some convenient scale of height is chosen.
- Points of the same height are joined together by a contour line.
- Often different ranges of height are coloured differently to emphasise the contrast; low-lying land is typically green and higher land brown.

We are so used to using and interpreting these contour maps that we don't need to think about it at all.

How does this help with the visualisation of functions of *3D* space? It may help to have a particular function in mind; let's say the temperature in a room:

$$T = f(x, y, z)$$

where:

T is the temperature at a point defined by
x : distance parallel to one wall
y : distance parallel to the wall perpendicular to the first wall and
z : is height from the floor, then
$f(x, y, z)$ is the function which, for each set of values of x, y and z, gives
    the numerical value of the temperature at that point.

Obviously, unless your central heating is *very* good, the temperature, T, in the whole space of the room will vary from point to point; there will be different values of T for different values of x, y and z. How do we represent this graphically?

Let's proceed as the geographers do with their 2D method:

- Some convenient scale of temperature is chosen.
- Points of the same temperature are joined together by a contour.

Now, we see the new thing; all the points of the same temperature in a (3D) room define a (2D) *surface* in the space of the 3D room and these *surface contours* will lie *inside* one another. This means that we can only '*see*' one of them at once. For example, if there is a heater in the middle of the room, then we would naturally expect to 'see'[1] a series of temperature contour surfaces roughly centred around the heater with lower temperatures the further from the source of heat.

The *theory* is clear enough; we can generate the contours but, unlike the geographer's contours, we cannot 'see' more than one at once.[2]

A.2   The Electron Distribution of the Lithium 2*s* Electron

Having set up the machinery with a familiar example, let's use a less familiar but simpler example to try the machinery out. The electron distribution of a 2*s* electron of the lithium atom is spherically symmetrical; whichever direction one proceeds outward from the nucleus the distribution is the same. What are the contours here and how do we display the information on a (2D) page?

- Because the distribution is spherically symmetrical, the contours are *spheres*. Convince yourself that this is true or the rest will be difficult.
- The spherical contours all lie inside one another centred on the lithium nucleus like the layers of an onion.
- Since the electron distribution does not fall away at a constant rate from the nucleus, the contours for equal changes in the distribution will not be equally spaced.

Here is a plot of the electron distribution as a function of distance from the nucleus:

---

[1]Loose terminology here: 'see' refers to the contour *diagram*, of course, not to the actual room!

[2]It is still a problem of dimensions because we cannot step outside 3D space as we can step outside the 2D space of the atlas page!

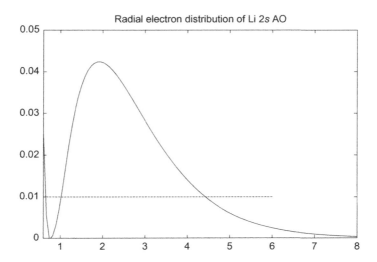

So, as we go out from the nucleus, the electron distribution:

- Falls very steeply at first;
- Touches zero (at about $R = 0.5$);
- Increases to a maximum (at about $R = 2$);
- And then falls away smoothly, tending to zero at large values of R (R greater than 8, say).

Let's see how we would represent this with contour surfaces in 3D space; in particular lets choose the value of 0.01 (the position of the dashed horizontal line). The horizontal line at 0.01 cuts the curve in three places:

(1) Close to the nucleus (remember the nucleus is at the origin of co-ordinates, $R = 0$).
(2) At about $R = 1$ on the upward part of the curve.
(3) Lastly, at about $R = 4.5$ on the final downward 'tail' of the curve.

This means that *at each* of these values of R (the distance from the nucleus) the electron distribution has the *same* value in all directions from the nucleus. So, there is a *spherical* surface (with the nucleus as centre) on which the distribution has the same value. In other words there are three concentric spherical *contours* around the nucleus where the value of the function is 0.01.

Obviously, we can pick any value of the distribution between zero and its maximum and generate spherical contours in this way. But how do we represent them on the 2D page? There are two kinds of approach:

- We can draw a sphere in perspective on the page in the usual way of graphic artists. This means we can only illustrate one contour on one diagram since they lie inside each other. Here is the outer contour:

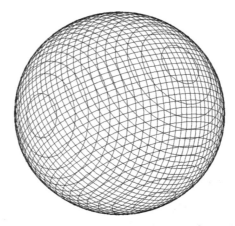

- We can take a 'slice' through the whole set in a plane containing the nucleus, usually (think of slicing an onion 'sideways' through the 'equator'). This will show us a section through the contours, we can see them all at once but only as *circles*, not as complete spheres. Here is the slice:

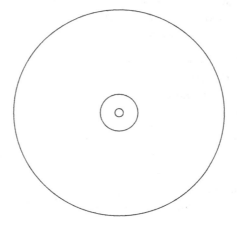

In order to appreciate the distribution fully, we really need to keep in mind *all three* of these representations:

(1) The initial graph showing how the distribution varies with distance from the nucleus;
(2) A 'typical' perspective of a 3D contour (in this case the important 'outer' contour where the horizontal line on the first graph cuts the 2s curve at about $R = 4.5$) and, finally;
(3) A 'slice' showing all three contours where the value of the 2s plot takes th value of 0.01 on the first graph to get a feel for how the three above representations fit together.

A.2.1 *How does this relate to the text-book 'orbitals'*

It is common in text-books to see simplified diagrams ('cartoons') of atomic orbitals or the associated electron distributions, and the 2s orbital would be a simple sphere. Obviously this correctly indicates the overall symmetry of the 3D function. It is less obvious what the contour actually represents. If we refer back to the original diagram on page 5 we can give an indication of the *size* of the distribution by choosing a contour somewhere on the downward tail. Say we decide to pick some minimum value on the right of the diagram ($R = 8$, say) and say that, after that point, the distribution has, to all intents and purposes, 'finished'. Then a contour at this value will indicate both the symmetry of the distribution and its 'size'. This kind of elementary information will prove a useful aid to visualising atomic and molecular electronic structures.

A.2.2 *What if the distribution is not spherical?*

If a distribution is not spherically symmetric the situation is more complicated because the contours are not spheres and their detailed shape really demands the use of the perspective drawing method.

If we take the next-simplest atomic orbital (AO) — one of the three 2p orbitals — then it is, perhaps, best to start with a perspective graph of a representative outer contour:

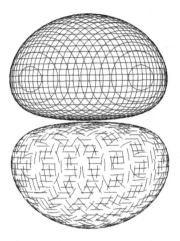

The fact that the 2*p* AO changes sign is indicated by full-line contours for the positive lobe and broken-line contours for the negative lobe. This diagram shows very graphically that *the value of the 2p function with distance (R) from the nucleus depends on the direction along which the direction is taken.* In particular there is a plane in which the value of the 2*p* AO is always zero: the plane which bisects the two lobes of the AO; here it is the horizontal line through the centre of the diagram. Cutting a slice through the 'middle' — i.e. passing through the lobes and through the nucleus — shows this property clearly, no contour crosses the horizontal line:

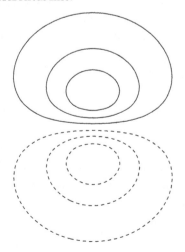

What is clear from these two diagrams is that a simple graph of value of a 2p AO plotted against distance from the nucleus is not as useful as the graph for the 2s AO since the *direction* of the line along which the value is plotted has to be specified.

This conclusion is strongly reinforced by looking at AOs (or, as we shall see, molecular orbitals (MOs)) of more complicated shapes (3d AOs for example) so, in general, only the two types of contour diagram will be used to illustrate the shape of orbitals in space.

Finally, note that it is either the perspective diagram of a typical outer orbital or just the outer contour from the 'slice' which is mimicked by the familiar cartoons to be found in standard chemistry texts.

Appendix B

Must We Use Quantum Theory?

All the calculations in the previous chapter and in the rest of this work were done using Schrödinger's mechanics (quantum theory). Yet some 'explanations' of chemical bonding make no reference to quantum theory or, indeed, any mechanics at all; using, for example, dots and crosses or other simple rules.

Contents

B.1 Connections to Laws of Nature

Although it is attractive to have laws and rules that are expressed in terms of ideas that chemists are familiar with, the science of chemistry can never be *independent* of the more basic laws of nature which govern the structure of matter. Similar situations arise in other sciences. Biology has its own laws, for example, but no biological system can be independent of chemistry any more than chemistry can be independent of physics. To take a very extreme example: no one doubts that a peacock is composed of molecules which are subject to the laws of chemistry, *but*, equally, no one would attempt to *derive* the form of the peacock's tail display from chemical laws.

However, the transformations involved when he produces this display must always *obey* the laws of chemistry.

It is in this spirit that one must always consider the relationship of any chemical laws and rules. They must always be at least *consistent* with the laws governing the behaviour of the matter of which chemical systems are composed: electrons and nuclei.

One of the most fundamental laws which governs the behaviour of charged particles like electrons and nuclei dates from the middle of the nineteenth century. In 1842 the Reverend S Earnshaw proved that a system composed of charged particles cannot be in equilibrium unless the particles are *in motion*; that is, electrons (and nuclei) are *never stationary*. This means, of course, that we must make the connection with some *laws* of motion for electrons.

B.2   Stable Molecules

We are very familiar with the idea of the 'bond energy' of a particular chemical bond. Roughly speaking, it is the energy 'tied up' in that bond when it is formed. Now, when the bond energy is actually *measured* and found to be (let's say) $250\,kJ\,mol^{-1}$, that measurement is carried out on billions and billions of molecules in the laboratory. We are used to simply assuming that each of these billions of bonds has *the same bond energy*: $250\,kJ\,mol^{-1}$ divided by the number of molecules in a mole (the Avagodro constant). That is, the bond energy is not some kind of *average* over those billions of bonds; they all have the *same* energy. The idea that the energy change during a chemical reaction is the same for each reaction *at the molecular level* involves the same reasoning. The 'energy of reaction' which we measure in the laboratory is not an average of many different energies, as each pair of molecules reacts with different individual bonds being broken and formed, but *in every case* the energy is the same. Think of what that means in terms of the motions of electrons:

- The molecules containing the bond in question are all colliding with each other and with the walls of the vessel and the atmosphere, and so exchanging energy during those collisions.
- But the bond energy is unaffected by this exchange of energy. Even when the collisions make the molecules vibrate, rotate and travel in different ways the bond energy remains unchanged;

• Which is just another way of saying that the motions of the electrons in a chemical bond and their energies are fixed at a constant value independently of the hard knocks they get in their normal environment.

Now our everyday experience is just the opposite. The planets and their moons going round the sun are similar in some ways to electrons around an atom or molecule, and yet they can change their motion and energy in any way and by any amount wheresoever; any collisions with asteroids or comets simply change the motion of a planet or moon by *some* amount. The energy of an orbiting body in the solar system is *not independent* of collisions and environmental effects.

The difference is that the motions and energies of large — 'everyday' — objects are subject to Newton's laws, but the energies and motions of very light, electrically charged objects are not. In order to get even the simplest facts of chemistry right — the constancy of bond energies — we are forced to use Schrödinger's mechanics (quantum theory). Only quantum theory[1] can give us a picture of nature in which energies of systems come in fixed, discrete amounts.

B.3   The Equipartition of Energy

The key idea here is that what we sense at an everyday level, as temperature is, at the atomic or molecular level, the translational motion (the kinetic energy) of a gas.

In the theory of the specific heats of gases, it is known from thermodynamic considerations that the amount of energy required to raise the temperature of, say, a mole of gas is dependent on the number of *degrees of freedom* in which molecules of the gas may move. The simplest case is that of an atomic gas such as helium, where the number of degrees of freedom is just three: the three independent degrees of translational freedom in which an atom may move. The natural assumption to make about what is happening at the atomic or molecular level is that the energy absorbed as heat is simply *shared equally* among the degrees of freedom of the atom or molecule. How could it be anything else? If one heats up helium the energy is equally distributed among the three translational motions; it does not

---

[1] That is what 'quantum' actually means: 'how much' in Latin.

suddenly start to try to move in some preferred direction! This apparently self-evident idea is the 'equipartition of energy principle'.

More complex molecules have *internal* degrees of freedom; in particular, they may rotate and vibrate. The general (non-linear) molecule will have just three degrees of rotational freedom (rotations about three axes in space) and a number of *vibrational* degrees of freedom which are more difficult to describe fully. Clearly a diatomic molecule has just one degree of vibrational freedom: the oscillations stretching and compressing the one bond. However, in general it is difficult to even *count* the number of independent ways a polyatomic molecule may vibrate. Imagine modelling the ethanol molecule as hard balls connected together by springs, and giving the model a good hard knock and counting the independent vibrations!

However, the total number of degrees of freedom can be obtained by a simple indirect method:

> If the number of atoms in a molecule is N then these N molecules have, between them, $3N$ degrees of (translational) freedom because each one has, like the helium atom, three degrees of freedom. Now, joining the atoms together to make the molecule does not change the number of degrees of freedom,[2] so a polyatomic molecule of N atoms has just $3N$ degrees of freedom.

Since we have seen that three of these degrees of freedom are the translation of the molecule *as a whole* and three are the rotations of the molecule as a whole, the remaining $(3N - 6)$ degrees of freedom must be the vibrations since there are no other possibilities.

Thus the specific heat of a gas should depend on its composition, in the sense that *the more atoms it contains, the more degrees of freedom can absorb energy, the more the energy is 'spread out', less of a given amount of energy goes into translation and so the more energy must be supplied to raise the temperature by one degree.* This is simply because the temperature of a gas is a measure of the amount of translational energy a molecule has. So, to take a simple example:

- The methane (CH_4) molecule contains five atoms, so has $3 \times 5 = 15$ degrees of freedom.
- If the energy supplied as heat is *shared equally* amongst the molecule's fifteen degrees of freedom, then a mole of methane will require *five times*

---

[2]This is not obvious but it is true!

as much heat to raise its temperature by a degree as a mole of helium which only has three degrees of freedom.

- This is not found experimentally. In fact, the specific heat of methane is nearer to *twice* that of helium.

What has gone wrong here? The *reasoning* seems perfect. It must be one or other of the assumptions made at the start that is wrong.

If we go back to the 'simple' case of a gas of helium it is easy to see what we have overlooked.

- *We think of* helium as a single atomic particle: three degrees of translational motion?
- Actually, at a lower level helium atoms consist of three particles, a nucleus and two electrons: 3×3 degrees of freedom?
- Lower still, since the nucleus contains four particles, two protons and two neutrons, so a helium atom contains six particles: 6×3 degrees of freedom?
- Even lower, each nuclear particle contains quarks etc.: there is no end to the complexity of the structure if we want to continue.

So the problem of the large number of degrees of freedom was there all the time; we had just inadvertently ignored it and, *just by good luck*, we got the right answer.

There is only one way out of the dilemma: it is not the *number* of degrees of freedom which determines the specific heat since this will always be larger than we (at the chemical level of things!) think, but it must be that many of the degrees of freedom which are actually there *are not involved in the 'sharing out' of the energy absorbed as heat.* That is to say, the absorption of heat (energy) is not continuous. It must be absorbed in different sized 'packets' (quanta) depending on the *type* of degree of freedom. In particular, heat energy (translation) does not come in large enough quanta to excite the motions associated with some degrees of freedom. In short, at the atomic and molecular level, the equipartition principle is wrong precisely because it is based on the false assumption of *continuous* energy sharing.

B.4 Quantum Summary

Using only information known since early in the twentieth century we have seen that there is no hope of being able to explain even the qualitative ideas of chemistry using the ideas of 'classical' (i.e. Newtonian) mechanics. This is not a question of detailed calculation of the motions of electrons but the most basic chemical ideas:

- The very idea of a bond energy or energy of reaction being the same fixed quantity at the molecular level is incompatible with the idea of the continuous transfer of energy between systems.
- Thinking carefully about apparently simple things like the kinetic theory of specific heats, shows that not only is energy exchanged in discrete amounts (quanta), there are differently-sized quanta for differently types of motion.

These are not the experimental facts which forced physicists to discard Newtonian mechanics at the molecular level, but they are things with which any chemist is now familiar. Hindsight is a wonderful thing.

Chapter 2

What We Know About Atoms and Molecules

The last chapter showed that we need to know something of the electron distributions in atoms and something of the way in which atoms are joined by chemical bonds before we can make a start on any theory of the electronic structure of molecules. This chapter lists what we should know and how that knowledge can be put to use in a theory of chemical bonding.

Contents

2.1 Atomic Electronic Structure

We have to start somewhere. What is the most sensible place to start with the development of a theory of valence? A theory of valence is not just about the electronic structure of molecules, but about how the electronic structure of molecules *relates* to that of the atoms of which molecules are composed. So let's start by reviewing the electron structure of *atoms*, which is assumed to be *known and familiar*.

2.1.1 *The hydrogen atom*

The equations of Schrödinger's mechanics are only soluble for the same kinds of system for which Newton's mechanics can be solved. The most relevant case for us is that of the interaction of two particles; in atomic theory this is just the hydrogen atom.[1] The energies and electron distributions resulting from the solution of the Schrödinger equation for the hydrogen atom provide the basis for the understanding of the electronic structure of most of the atoms of the periodic table, and so it is worthwhile looking at them again in some detail:

Energy Levels The energy which a hydrogen atom may take up is *quantised*; hydrogen atoms may only have *certain, fixed* amounts of energy. The formula for these allowed *energy levels* (as amounts of quantised energy are usually called) is

$$E_n = -\frac{R}{n^2} \qquad (2.1)$$

where:

- E_n is the value of the nth energy level.
- n is a positive integer (a positive whole number; $n = 1, 2, 3, \ldots$)
- R is a positive constant — called the Rydberg constant — composed of fundamental constants like the charge on the electron and on the proton.

Note that the *sign* of the expression on the left of equation (2.1) is *negative*; this is simply a sign convention. E_n is the energy of the hydrogen atom with respect to a zero of energy, which is that of an electron and a proton separated. Of course, the energy of the atom is *lower* than that of its components, otherwise it would not form. Therefore it is negative (lower than zero). This prompts an obvious question:

> What about a hydrogen atom with *positive* amounts of energy?

---

[1]Strictly, any one-electron ion as well: He^+, Li^{2+} etc.

The answer is just as obvious:

> A 'hydrogen atom' with more energy than a separate
> electron and proton is not an atom; it is just an electron
> and a proton whizzing about separately.

Electron Distribution The solution of Schrödinger's equation for the
hydrogen atom, unlike the solution of Newton's equation for the
moon going round the earth, does *not* give the *path* of the electron in
'orbit' around the nucleus. However, this solution does give a different
description of the motion:

- It shows that the electron is held 'captive' by the pull of the nucleus.
- It gives a quantitative measure of the probability where the electron
  will be. This probability is just a function of (3D) space; it does
  not change with time.

The solutions of the Schrödinger equation are called *orbitals*. The
word is chosen because of its connection with the idea of 'orbit' which
is the Newtonian description of the motion of the two-interacting-
body problem. The orbitals themselves do not give the probabilities
directly; it is the *squares* of these functions of 3D space which tell us
the relative probabilities of where the electron in the hydrogen atom
will be.

The explicit functional forms of these orbitals and their nomenclature are familiar and will be simply reviewed here. The general
notation for an orbital is

$$\text{Integer Letter}_{\text{subscript}} \tag{2.2}$$

where:

- 'Integer' is the n of equation (2.1), i.e. the integer which fixes the
  energy of the electron described by this orbital.
- 'Letter' is just a code letter which describes the general shape (in
  3D space) of the main contours of the orbital.
- 'Subscript' is again a code to describe the orientation and symme-
  try of the shape defined by 'Letter' in 3D space.
- Particular examples are very familiar:
  Integer $= 1$; Letter $= s$; subscript $=$ nothing; gives $1s$
  Integer $= 2$; Letter $= p$; subscript $= z$; gives $2p_z$
  Integer $= 3$; Letter $= d$; subscript $= xy$; gives $3d_{xy}$
  etc. etc.

We need to look carefully at the interpretation of these orbitals. This is done in Appendix C at the end of this chapter.

> These orbitals for the hydrogen atom (and, as we shall see, for other atoms) are called atomic orbitals, which is universally abbreviated to AOs, one talks about, for example, the $3d_{xy}$ AO of an atom.

Terminology There is some standard terminology associated with these energy levels and the corresponding AOs. Rather than describing the lowest state of the hydrogen atom as

- 'An electron bound to a proton with energy $E_1 = -R/1^2$ and with electron probability given by the square of the $1s$ AO function' one simply says;
- 'An electron occupying a $1s$ AO'.

This terminology is compact and convenient and is used throughout chemistry. In the same spirit, if the electron of hydrogen is excited to a higher level one says that the electron is 'excited *from* the $1s$ AO *into* (say) the $2p_x$ AO'. The convenience of this terminology has one rather unfortunate effect. It tends to encourage the mistaken impression that:

> The electrons occupy the AOs in much the same way as rabbits occupy hutches; that the orbitals exist 'out there' in the world just waiting to be occupied by electrons whereas, as we explain in Appendix C, this is not the case.

But, so long as these limitations are kept in mind, this terminology is always useful.

A Feature of the Energy Levels The fact that the energy levels of hydrogen and, in fact, other atoms are described by a formula like equation (2.1), which makes the energy depend *inversely* on the square of an integer, means that, as the allowed energy goes up, the levels *get closer and closer together*:

$(1/1^2 - 1/2^2) = 0.25$ is larger than $(1/2^2 - 1/3^2) = 0.13\dot{8}$ and so on.

This has some consequences for chemistry, as we shall see later.

2.1.2 Many-electron atoms

The outlook for the calculation of the electronic structure of many-electron atoms does not look too optimistic. The Schrödinger equation is not soluble, so we cannot get expressions like equation (2.1) for their energies. Think about what is involved in calculating the energies and electron distributions in, for example, the oxygen atom with a nucleus (charge $+8$) and eight negatively-charged electrons:

- The electrons are strongly attracted to the nucleus; they have charges of opposite sign.
- But the electrons all repel one another because they have identical charges.

Each electron's distribution depends on the distribution of all the others. This looks like a Catch-22. How can we calculate the distribution of any electron if, to do so, we need the distribution of all the others, and the distribution of these others, in turn, depends on the distribution of the electron we started with?

Fortunately, there is a solution based on the idea of *self-consistency*. What is done is to:

(1) *Guess* the distribution of all but one of the electrons
(2) and use these guessed distributions to calculate the distribution of the remaining one.
(3) Then, armed with this calculated distribution we can use this and all but one of the guessed distributions to calculate the distribution of another electron
(4) and continue in this way replacing the guessed distributions with the calculated ones until the (guessed or calculated) distributions stop changing.
(5) Then the whole set of distributions for the atom is, at least, self-consistent.

But where are the guesses in step 1 of this process to come from?

This is where the AOs of the hydrogen atom come in. The argument goes like this:

> The hydrogen atom is, certainly, not a 'typical' atom, but it is not *very* unusual. So, perhaps, the electron distributions

in many-electron atoms are not *very* different from those of the hydrogen atom. That is, why not use the (squares of the) AOs as guesses for the electron distributions in the above process?

This sounds perfectly reasonable except for one thing:

The lowest state of the hydrogen atom has one electron occupying the $n = 1$ ($1s$) level. Surely the other atoms will have their electrons in the lowest possible level. So why do not all the other atoms simply have all their electrons in the lowest ($1s$) level?

If this were true then, for example, the helium atom would have two electrons in the $n = 1$ level, lithium would have three, beryllium four, and uranium would have 92 electrons in the $n = 1$ level. In other words, there would be no periodic table because the electronic structure of *all* atoms would just differ by how many electrons were in the $n = 1$ level. There would just be a *gradual* change in atomic properties with increasing atomic number instead of what we do see: sudden changes in atomic properties as the atomic number changes (think of the difference in the properties of the neon atom with 10 electrons and those of sodium with 11, just one more).

This is a very serious problem and there is no answer to it to be had from the Schrödinger equation; indeed if we carry out calculations on the electronic structure of, for example the fluorine atom, we do get this structure as the lowest state with an energy very different from (much lower than) the experimental value. What is to be done?

2.1.3   *The Pauli principle*

In trying to understand the electronic structure of many-electron atoms we have stumbled across an important omission in the analogy between Newton's mechanics, (which is applicable to large or everyday objects) and Schrödinger's mechanics, which is relevant at the sub-atomic level. This constraint on the distribution of electrons proves to be the third vital piece of equipment which we need to develop a theory of the laws of valence; the way in which atoms combine to form molecules. The other two items are: Coulomb's law of electrostatics and Schrödinger's mechanics.

There are several ways in which the Pauli principle may be introduced but, basically, it is a new fact which must simply be acknowledged as a law of nature and incorporated into our theories. For the moment, it is sufficient to say:

2.1.3.1 *Statement of the Pauli principle*

> When a many-electron system is described by electrons occupying orbitals, no more than two electrons may occupy any one orbital.

The term 'orbital' rather than 'atomic orbital' has been used deliberately, since the principle is applicable to orbitals in any kind of matter: atoms, molecules, solids.

2.2 Using the Atomic Energy-Level Scheme

We now have the information necessary to be able to extend the energy-level system of the hydrogen atom to many-electron atoms. First of all, it is very useful to display the energy levels as a diagram. For the hydrogen atom this is a very simple task; we simply start at the lowest (1s) level as an origin and draw the other levels upwards with the (decreasing) gaps between them:[2]

AO Energy Levels for the Hydrogen Atom

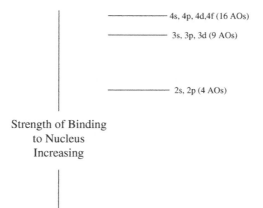

---

[2]Remember equation (2.1), all energies of bound electrons are *negative* by convention, so the lower the level on these diagrams, the more tightly bound an electron is to the nucleus in that level.

The first thing to notice is that, since the energies of the H atom depend only on the integer n, all the AOs with the same n (a total of n^2) have the same energy; they are, in the standard terminology, *degenerate levels*. So, if we would like to show the AO energies in more detail on the diagram we can simply 'explode' the diagram sideways showing each *type* of AO (ns, np, nd ...) separately:

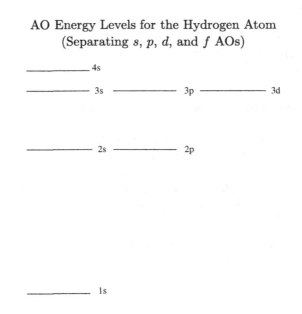

AO Energy Levels for the Hydrogen Atom
(Separating s, p, d, and f AOs)

If it is useful to give *each* AO a place on the picture (e.g., all three np AOs) this process can be carried further, making the diagram wider and wider. It is usually sufficient to use the second diagram above, as we shall see.

In the case of many-electron atoms the diagram is similar to the one above but, for reasons to be discussed later, the degeneracy of all levels with the same n is removed; for a given n the ns, np, nd sets of AOs no longer have the same energy:[3]

---

[3]The three np AOs (np_x, np_y, and np_z) are simply labelled x, y and z, and the five nd AOs (nd_{xy}, nd_{yz}, nd_{xz}, nd_{z^2}, $nd_{x^2-y^2}$,) are labelled by their subscripts in the same way in the diagram.

AO Energy Levels for Many-Electron Atoms
(Showing Each AO separately)

—————————— 4s

—————————— 3s $\underset{x}{—}\ \underset{y}{—}\ \underset{z}{—}$ 3p $\underset{xy}{—}\ \underset{yz}{—}\ \underset{xz}{—}\ \underset{zz}{—}\ \underset{xx\text{-}yy}{—}$ 3d

$\underset{x}{—}\ \underset{y}{—}\ \underset{z}{—}$ 2p

—————————— 2s

—————————— 1s

The central question now is 'can we use this diagram to explain the electronic structure of atoms and, if so, how?' In particular, can we explain the structure of the periodic table?

We said in a previous section that the electrons would, presumably, take up a distribution which was associated with the lowest energy available to them. The Pauli principle places a severe restriction on what is meant by the phrase 'lowest energy available', since only two electrons may occupy a given AO after helium the 1s AO is 'full'. Thus, an additional electron has to go into the next-lowest level: the 2s AO. Obviously, beryllium follows the same pattern and its electronic structure is $1s^2 2s^2$.

The boron atom apparently poses a problem: there are three 2p AOs, which one does the 5th electron occupy? The three 2p AOs all have the same energy for a very simple reason: they are exactly alike except for their orientation in space. Now, we have orientated them with respect to the three perpendicular directions in space, labelled by the three Cartesian axes: x, y and z. That's fine for *us* to distinguish between them but an electron does not 'know' anything about the directions we have chosen, so a 2p AO in *any* direction in space will do. In short, we simple say for boron 'an electron in *a* 2p AO'. Continuing in this way gives unique structures for the next atom, carbon (atomic number, Z, of 6). In this case there are two distinct possibilities:

(1) The two electrons which must be placed in the set of three $2p$ AOs can be placed in the same $2p$, AO i.e. any one of $2p_x$, $2p_y$, $2p_z$, since all three have the same energy.
(2) The electrons can be placed in two separate $2p$ AOs, i.e. one each in any pair of the three $2p$ AOs.

Using the above diagram there is nothing to choose on energy grounds between these two possibilities since all three $2p$ AOs have the same energy and, of course, neither of the two structures violate the Pauli principle.

Although we can ignore the actual orientation in space of the *set* of $2p$ AOs, the two above structures are distinct; one has two electrons in a single $2p$ AO oriented in any direction in space while the other has one in each of two $2p$ AOs which are in mutually perpendicular directions. The absolute direction of the pair is unimportant, but their *relative* orientation — with respect to each other — is fixed.

This is a new problem and requires a new solution. In using an energy-level scheme based on that of the hydrogen atom, and changing it slightly to accommodate the fact that not all the energy levels for the same value of n are the same, we have skipped an important contribution to electronic structure which we have met again in thinking about the electronic structure of the carbon atom.

> The difference between the energy-level diagram for hydrogen and the one for many-electron atoms has not been explained, and we are unable to choose between the two structures for the electronic structure of the carbon atom. The reason is obvious: what is the biggest difference between the H atom and *all* other atoms? All atoms other than hydrogen contain more than one electron, electrons are charged and so repel one another — we have omitted to include the repulsions between electrons in our considerations! This, in spite of stressing the key rôle of Coulomb's law in understanding the structure of atoms and molecules.

Postponing for the moment the first problem and keeping to the problem in hand, if we think for a moment about the two possible structures for the carbon atom, the choice between them is easily made by including electron repulsion. Repulsion between electrons is *positive*, i.e. destabilising, energy-raising quantity. If two electrons occupy the same AO they are clearly, on average, closer together than if they occupy two different AOs which are mutually perpendicular. The destabilising repulsion is larger in the configuration which has two electrons in the same AO.

This provides the only piece of information which we need to complete the description of the periodic table using the energy-level diagram. The hierarchy of rules for occupation of AOs is, then:

(1) Place each electron in the AO with the lowest energy.
(2) The Pauli principle: one AO may only contain a maximum of two electrons.
(3) In occupying AOs which are degenerate — have the same energy — fill up the degenerate set singly before doubly occupying any AO in the set.

It is completely clear here that it is the Pauli principle which causes the appearance of similar electronic structures, as the number of electrons increases. For otherwise, as we said in Section 2.1.2, all the electrons would pile into the $1s$ AO until their mutual repulsions became so great that occupation of higher levels was the only way to relieve this destabilising effect.

There is a simple standard notation for the electronic structure of atoms obtained in this way:

> One writes an electron configuration (for example, for the oxygen atom) as:
>
> $$1s^2 2s^2 2p^4$$
>
> and, in general, a string of AO symbols with superscripts indicating the number of electrons contained in a given *type* of AO; using 'type' to mean a set of AOs which all have the same energy, such as $2p$, $3d$ or $1s$.

The example above, which contains the notation $2p^4$, does *not* mean that the Pauli principle is violated by a single AO containing four electrons, it means that the *set* of three $2p$ AOs ($2p_x$, $2p_y$, $2p_z$) contains a total of four electrons distributed among the three in any way which satisfies the Pauli principle and our electron-repulsion rule. Similarly, in the electronic structure of nickel ($1s^2 2s^2 2p^6 3s^2 3p^6 3d^8 4s^2$) the term $3d^8$ means, not that an AO contains eight electrons, but that the *set* of five 3d AOs contains eight electrons.[4]

Although this set of rules[5] has been derived for the electronic structure of atoms, a little thought shows that it all applies to any set of orbitals whatsoever, so we shall be able to use the same rules without further ado when we come to molecular electronic structure.

2.2.1   *Current summary for atoms*

This completes the apparatus one needs to picture (and to calculate) the electronic structure of atoms. It is simply assumed that the hydrogen-type AOs are to be used for the description of the AOs of the many-electron atoms. More importantly, it provides the background and justification for the use of the familiar AOs in the *qualitative* description of the electronic structure of atoms. Since one of the approaches used in the description of molecules is the fact that molecules may be considered to be made of atoms, we can expect to be using this apparatus in our considerations of the electronic structure of molecules. To summarise then:

- The electronic structure of atoms is determined by:
  - (1) The law of force operating between charged particles: Coloumb's law,
  - (2) Schrödinger's mechanics,
  - (3) The Pauli principle.

- The AOs obtained from the solutions of the Schrödinger equation for the hydrogen atom can be used to describe the structure of many-electron atoms.
- The Pauli principle, restricting the occupancy of AOs, generates the periodicity of properties of the atoms.

---

[4]A little thought shows the sense of this compact notation since there are often several ways of distributing the electrons amongst the AOs of a given type which all have the same energy.

[5]There is a fourth rule which relates to the orientation of the spins of the electrons in a degenerate set of orbitals, but since electron spin effects are of little importance in the theory of valence, this is not given here. See Appendix H for elaboration of this point.

2.3 Empirical Chemistry

For over a hundred years molecules have been represented in the now-familiar way as a set of atomic symbols linked by lines. Each of these lines conventionally represents a chemical bond. During that time these lines have been interpreted in many ways, from the early days when the lines simply indicated which atoms were joined together, to the present-day quantum-mechanical view of the bonding. But each of the successive interpretations of the lines has not *completely* replaced the earlier versions, and there is a kind of historical 'husk' of outdated ideas surrounding the modern theory of the chemical bond. So we have to try to look at the idea of the chemical bond with fresh eyes.

The first theory of the chemical bond which said something about the 'mechanism' of bonding — the Lewis 'dots and crosses' theory — involved the idea that the chemical bond was associated with the sharing of electrons between atoms, each of the lines being thought of as representing a pair of electrons. Modern theories still retain this central idea, but are now able to *explain* the mechanism and details of this electron sharing.

It is important to have a clear idea of what is meant by 'explanation' in science and, in particular, in the theory of the chemical bond. It does *not* mean simply:

- A convenient way to remember some facts: e.g. 'the valency of carbon is 4';
- A rule of thumb to get the numbers right: e.g. 'octet (8-electron) or 18-electron rules';
- A pictorial way of representing some ideas: e.g. 'dots and crosses';
- A numerical rule which (sometimes) gives the right answers: e.g. 'counting bonding and anti-bonding electrons',

although a good explanation may well do some or all of these things. What it *does* mean is, for the time being anyway,

> A way of describing the phenomenon of chemical bond formation, which uses the properties of the particles of which the system is composed (electrons and nuclei), and the laws of interaction amongst these particles.

Notice the fact that, as emphasised in Chapter 1, this does not mean that chemistry is just the dynamics of electrons and nuclei. We must make a *chemical* choice of what the 'system' to study is.

Going back to the start, a chemical bond is simply represented by a line drawn between two chemical symbols between atoms. This symbolism is conveniently ambiguous since:

- The line may simply be a convenient way of indicating 'what is joined to what' in the molecule, or it may indicate some details of the structure of the molecule.
- The atomic symbols may be thought of as representing the *atoms* or the *nuclei* of those atoms.

The approach to be used here is to look at the chemical structural formula — atomic symbols joined by lines — and interpret *all* these parts of the diagram — atomic symbols *and* lines — in terms of electronic structures. This will remove the above ambiguities.

- If we think of the simplest molecule H_2 then, since the electron distribution is *all* associated with the *line* joining the atoms, rather than with the atoms themselves, the two 'H' symbols in the structural formula H–H are surely the nuclei rather than the atoms.
- If, on the other hand, we look at the iodine molecule, I_2, the line represents only two of the 106 electrons in the molecule. Here in the structural formula I–I it is more sensible to think of the two 'I' symbols as representing the atoms rather than the nuclei.

Obviously the molecule H–I represents an intermediate case. What is obvious from these rather elementary considerations is that:

> The lines representing bonds in the conventional structural formulae for molecules are *not* simply an indication of 'what is joined to what' in the molecule. They are shorthand for the important non-atomic (i.e. *molecular*) components of the electronic structure of that molecule: the electrons which are responsible for the bonds that those lines indicate.

If we are thinking of ways to approach an understanding of the structure of molecules, then it is clear that we are going to need something other than the simple AOs — the functions which describe the electron distributions in the *separate* atoms — to get a handle on how the electrons are distributed in those innocent-looking links in the structural formulae.

Presumably, the vast majority of the 104 electrons in the iodine atoms which are not involved in the chemical bond can be more or less accurately described by the AOs of the iodine atom precisely because they are not involved in the changes associated with bond formation. But the two electrons involved in the bond must take up a distribution different from those of the separate iodine atoms, and so need to be thought about more carefully. In fact, of course, if the distributions of two of the 'outer' electrons in the iodine atoms are changed by molecule formation, then this will have repercussions for any other outer electrons since the repulsions amongst these electrons will be disturbed by that change. This will show itself as the reorganisation of the 'non-bonding' electrons of the atoms into lone pairs. This reorganisation, in turn, may imply further bonding possibilities: the I_2 molecule may bond to other systems. We shall have to look at these complications later particularly in the context of transition metal chemistry.

On the basis of this tiny amount of data[6] it is proposed to make the provisional assumption that:

> The atomic symbols in a conventional structural formula will be taken to mean the so-called 'atomic cores', that is, the atomic nucleus *plus* all those electrons which are not involved in any electron-sharing with other centres in the molecule.[7]

This provisional definition will have to be revisited later, since it does not make the important distinction between the tightly-bound 'inner' electrons (the real cores) and any electrons in the 'outer' part of the atom which are not involved in chemical bonds (the lone pairs).

Further, we may assume, again provisionally:

> The lines between atomic symbols in a structural diagram of a molecule each represent a pair of electrons which are shared between the two centres indicated by the atomic symbols.

---

[6] The reader can, no doubt, supply all the additional data needed here!

[7] This idea of 'atomic cores' will be used later to put the controversial idea of 'oxidation number' on a firm theoretical basis. Of course, the hydrogen atom has only one electron and its atomic core is just the nucleus.

And, finally, that:

> The separation of the electrons in a molecule[8] into 'atomic
> cores' and 'bonds' will provide the division we are seeking
> into the 'environment-insensitive' substructures which will
> enable their energies and distributions to be described, both
> qualitatively and quantitatively.

What we need now is a method of approach to the description of the electronic structures of these two types of molecular electronic substructure.

2.4   Assignment for Chapter 2

The familiar diagram which gives the energy levels available to electrons in the first 30 elements of the periodic table differs from the corresponding energy-level diagram for the hydrogen atom, since not all the levels for a given n have the same energy. In general the rule is:

$$E_{ns} < E_{np} < E_{nd} < E_{nf},$$

where E_{ns}, E_{np}, E_{nd} and E_{nf} are the energies of the s, p, d and f AOs for a given n (where these are relevant).

Problems:

(1) Why is this so? What is the most obvious difference between the hydrogen atom and all other atoms? Discuss the various electrostatic interactions in the atom to try to explain this important fact. Concentrate on, for example, the different *shapes* of the $2s$ and, $2p$ AOs in the boron atom, and how electrons having these different distributions might interact differently with the attractive pull of the nucleus, and the repulsions between these AOs and the 1s core electrons. Explain why, as n increases, the nature of the periodic table changes — explain the appearance of the transition metals and the rare earth elements.

(2) In some text-books you will find the third rule in Section 2.1.2 replaced by a rule based on what is called Hund's rule,[9] which states that electrons in degenerate (same-energy) orbitals 'prefer' to have their

---

[8]Containing no lone pairs, remember.

[9]After a famous physicist Friedrich Hund, there are other Hund's rules but they relate to the details of atomic spectra.

spins parallel. 'Electron spin' is a name given to the fact that electrons behave as magnets.

What is the preferred orientation of two magnets: parallel (north pole to north pole, south to south) or anti-parallel (north to south, south to north)? In fact, electrons in degenerate orbitals *do* prefer to have parallel spins. Explain this strange discrepancy — you will need to discuss this with your tutor.

Look at Appendix H.2 to get a handle on this last problem

(3) Do you find the idea that the lines in the structural diagrams of molecules represent something 'real' acceptable; that is, these lines are not just indications of which atoms are joined but represent a definite electronic structure — usually a two-electron bond? This idea will not always be useful, ask your inorganic tutor to give some examples.

Appendix C

The Interpretation of Orbitals

Let's say, once and for all, what an (atomic) orbital is and how it is to be interpreted. This means thinking a little about probabilities.

Contents

C.1 What is an Orbital?

Before looking in detail at the interpretation of orbitals we ought to give some preliminary answer to the question 'what is an orbital?' The familiar cartoons of atomic orbitals are everywhere in chemistry texts, and these cartoons are generally associated with the explanation that 'the square of an orbital is the probability distribution for the presence of an electron'.[1]

It is common to ask questions like:

- Do orbitals exist 'out there' in the real world?
- Is an orbital what you would see if the atom were large enough to see?

The answer to both these questions is 'no', and the explanation why this is so provides a useful way of approaching the main task of the interpretation of orbitals.

---

[1] More often the phrase is 'the square of an orbital is the probability of *finding* an electron', but we can deal with this mistake later.

The square of an orbital is a 'probability distribution' and, like any other probability distribution of a more familiar kind, on a human scale, this one has a strict interpretation. If we look at two probability distributions for familiar, macroscopic, situations it helps to justify the answers to these two questions:

(1) Let's say that accident statistics are sufficient to be able to create a good approximation to the probability distribution of road traffic accidents (RTAs) on a stretch of motorway. We might expect larger numbers of accidents close to junctions or other kinds of changes in the smooth path of the road, for example, so that the probability distribution as a (one-dimensional) function of distance might look something like this:

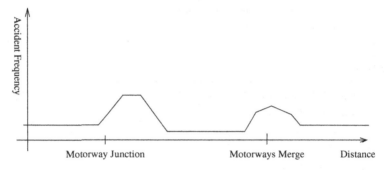

(2) The probability distribution for the position of a simple pendulum is very easy to calculate as a function of angle, from the assumption that it executes simple harmonic motion and looks like:

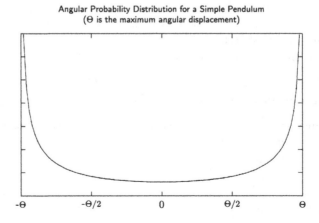

Angular Probability Distribution for a Simple Pendulum
(Θ is the maximum angular displacement)

The appearance of this distribution is easy to justify. The probability is inversely proportional to the velocity of the pendulum bob. It has a higher probability of being at the ends of the swing, where it stops momentarily and reverses than in the centre of its swing, where it is travelling at its maximum velocity.

Now we can see that *neither of these distributions* occurs 'out there' in the real world.

(1) If one looks at a length of motorway all one sees are 'accidents' or 'no accidents'. There is no sign of the probability distribution *function* anywhere on or near the motorway — it is a mental construct, it exists in our heads or on paper or in our computers.

(2) Likewise, if one makes random observations of the position of a pendulum at a point in its swing, what one sees is either 'a pendulum' or 'no pendulum'. Again, there is no sign of the probability distribution *function* at or near the pendulum — it also is in our minds, on paper or whatever. It is a summary of the pendulum's behaviour which we have generated to help us understand the way a pendulum moves and is distributed in space.

Notice that the non-existence in the outside world of either of these probability distributions does *not* imply anything mysterious about road traffic accidents or pendulums, and it certainly does not imply that either RTAs or pendulums are spread out in space like waves or whatever. If we wish to verify the information contained in probability distributions like these we have to perform many observations and, say, plot the results (e.g. numbers of 'pendulum' or 'no pendulum' results) against position.

Now we can make sense of the answers to the two questions about orbitals at the start of this section:

> If we 'look' at, for example, a hydrogen atom with an electron in the $2p$ atomic orbital,[2] what we see is either 'an electron' or 'no electron' at a particular place. The probability distribution — the square of the $2p$ AO function — is not there; it is in our minds as a convenient summary of many, many observations.

Now let's look more carefully at orbitals and what they are.

---

[2] Using 'look' to mean something like 'get all the experimental kit together to detect an electron at a particular place'.

C.2   Orbitals: Atomic and Molecular

An orbital (of whatever kind: AO or MO) is a function[3] of ordinary (3D) space which, when squared, gives the probability distribution for the presence of an electron in space. The electron is a *particle*. It is not 'spread out' in space any more than an individual road traffic accident is spread out over the motorway. An electron is or is not at a particular place in space in just the same way as a pendulum is or is not at a particular angle in its swing. The way that one can experimentally verify whether quantum theory gives a good description of nature or not has, therefore, to be *statistical*; we must perform many, many measurements of electron position and compare these measurements with the probabilities predicted by squaring the orbital.

This is not so tedious as it might seem. In practice the sorts of experiments which are carried out to measure 'electron density' (e.g. diffraction experiments) are, in fact, carried out on small samples of material, but these small samples contain billions and billions of atoms or molecules,[4] so the conditions are already met.

---

[3]Strictly speaking, an orbital is a function *which is a solution of some Schrödinger equation that has these properties.*

[4]Think of the size of Avogadro's number and calculate roughly how many examples of a given chemical bond there will be in one milligram of any organic molecule of about 20 atoms.

Chapter 3

A Strategy for Electronic Structure

Now is the time to put together the general ideas from Chapter 1 about choosing the most appropriate overall structure for a molecule and the ideas we have reviewed in the previous chapter about atomic and molecular electronic structure.

Contents

3.1 Review

For the moment, continue to think about a molecule which has no lone pairs, i.e. there are no electrons which are not involved in bonding with similar energies to the electrons involved in the chemical bonds. The electrons in such a molecule fall into two types:

(1) The electrons in the atomic cores; the ones which are so tightly held by the nucleus of one atom that their distribution is not appreciably changed by the presence of other atoms in the molecule. This means, for example, in most simple organic molecules, the electrons in the $1s$ AO.

(2) The electrons in the chemical bonds; those electrons whose distribution in the molecule is very different from their distribution in the separate atoms.

This is, at least, one sensible way to divide up the complex overall electronic structure of even this, relatively simple, molecule into relatively independent parts. The problem now is how to apply the known laws of interaction between electrons and quantum theory to give a description of the motions, energies and distributions of those electrons.

- It is obvious how to describe the first set of electrons because, by our very assumption, their distributions are the same as they were in the separate atoms. So the electrons in the atomic cores occupy the AOs of their 'own' atom. If the atom is a carbon atom in (say) CH_4, its atomic core is the nucleus and just two electrons occupying the $1s$ AO of that carbon atom *which we assume to be known*. If the atom is a silicon atom in (say) SiH_4, its core is the nucleus together with the ten electrons in the structure $1s^2 2s^2 2p^6$ of the silicon atom.
- In both of these cases, there are four electrons from the atom which are involved in the bonds of the molecule and have distributions different from their AOs in the separate atom. What *about* the electrons in the chemical bonds? They, after all, are the 'interesting ones' in our investigations.

This looks like a tricky problem. Think about what is involved:

- We want to describe the motion of two electrons whose motion and distributions are influenced by the attraction of two atomic cores simultaneously. That is, the net attraction due to:
  (1) The attraction between the nuclei and these two electrons.
  (2) The repulsions between any tightly-bound electrons in the atomic cores and the bond-pair of electrons.

  Plus, of course, the smaller effect of the repulsions between the two electrons in the bond.

In order to get a handle on this problem we have to think about how a similar, but less complicated, problem has been solved: the electronic structure of many-electron atoms.

In a many-electron atom like carbon or silicon the distribution of the electrons is governed by the attraction of each electron to the (single) nucleus and all the *mutual* repulsions amongst the electrons. So in the carbon atom, for example, the motion of any one of the electrons is determined by its interactions with all the other six particles (one nucleus and five other electrons). But, in fact, we are no longer deterred by this apparently complex problem.[1]

> What we did in this case was to *make a model* of the overall structure *based on our physical and chemical knowledge of the behaviour of atoms*. It is known what the structure of the hydrogen atom is[2] and, *crucially*, the properties of atoms can be modelled by using the main features of the hydrogen-atom structure. In a few words, we used the AOs of the hydrogen atom to *model* the structure of many-electron atoms.

Of course, the *details* of the AOs differ from atom to atom but the overall idea gives excellent agreement with a mass of experimental data.

In saying above that the atomic cores may be described by the AOs of the separate atoms, we are taking this model over into a new environment and saying that the electron distributions of the atomic cores are (i) capable of being described by an AO model and (ii) very similar to their distributions in the isolated atoms. What remains, then, is to make a suitable overall model of the electronic structure of a pair of electrons 'shared' between two atoms: the electrons in a chemical bond. The only way forward is to think about what we know about atoms and molecules.

We will have to look back at our results of Section 1.4 on page 8, where we looked at what happens to the atomic electron distributions when two atoms approached each other. What was found there was that the electron distributions were *polarised* by the presence of another atom and, although that is not surprising, what has not yet been done is to find a method of *describing* this process of polarisation. What we need to do now is think about how to describe:

(1) *Polarised* AOs;
(2) The way in which distributions are changed *when the electrons undergo further change to form a chemical bond*.

Since the hydrogen molecule H_2 is a special case, in the sense that the electrons in its $1s$ AO are relatively tightly-bound and show little

---

[1]The dynamical equations are, in fact, insoluble in both classical and quantum mechanics.

[2]The relevant equations *can* be solved.

polarisation on bond formation, the more typical, but slightly more complicated, lithium hydride molecule is used as the principal example.

3.2   Lithium Hydride Again

If we look more carefully at the (polarised) AO distributions of the lithium hydride molecule which we looked at on page 11:

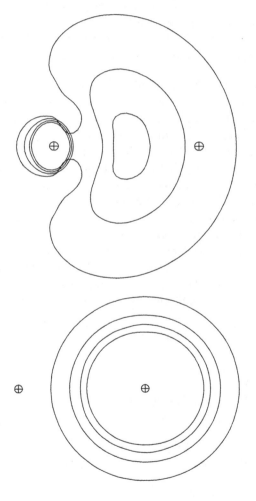

we can, perhaps, see a way forward.

3.2.1 *Polarisation and hybrid AOs*

So far, we have stressed that the probability distribution of an electron is the square of an orbital. Let's now look at the converse: finding an orbital which, when squared, generates an electron probability distribution. In particular, how do we find an (atomic) orbital which, when squared, will generate an electron distribution like the one in the diagram above; the polarised electron distribution of the valence electron on the lithium atom as another atom approaches?

First of all, what 'materials' have we got to hand? As usual, in the description of the electronic structure of atoms, we must think of AOs, the standard building blocks. The AOs $2s$ and $2p$ (three of them, remember: $2p_x$, $2p_y$ and $2p_z$) all have the same energy in the hydrogen atom and similar energies in many-electron atoms. All the other AOs have energies different from these four. The appearance of the AOs is probably familiar. The contours of the $2s$ AO are spheres and those of the three $2p$ orbitals — the same as each other except for orientation in space — are the 'pinched-in-sphere' or 'dumbbell' shape:

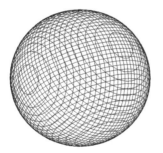 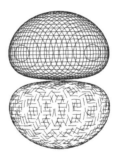

Obviously, the $2s$ is the original unpolarised AO occupied in the ground state of the lithium atom, while the electron distribution, which is the square of one of the $2p$ AOs, is very polarised along the axis which is used to label it $2p_x$, $2p_y$ or $2p_z$. Comparing these diagrams with the lithium diagram suggests an approach. Noting that the $2p$ AO consists of two 'lobes' of opposite sign while the $2s$ AO is a sphere of single sign, we can see that the $2s$ is not polarised at all while the $2p$ is too polarised. So perhaps a *linear combination* of the two will suit our purpose. We can express this first in words, then in symbols. For some choice of numbers a and b:

Polarised AO = (Part of $2s$ AO) + (Part of $2p_z$ AO)

= (A Number) $\times$ ($2s$ AO) + (A Number) $\times$ ($2p_z$ AO)

i.e. $\phi(x, y, z)$ = $a2s(x, y, z) + b2p_z(x, y, z)$,

where ϕ is an atomic[3] orbital which will, when squared, generate an asymmetric electron distribution polarised in the z-direction. Here it is, for a particular choice of the combination coefficients a and b: $a = 1/2$ and $b = \sqrt{3}/2$:

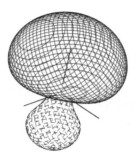

The diagram has been tilted slightly to give a better view, with the z-axis not vertical but along the line through the middle of the large lobe. This choice of the coefficients a and b gives the so-called sp^3 hybrid, which is commonly assumed to be the one involved in, for example, C–H and C–C single bonds.

The *atomic* orbitals generated in this way are obviously neither 'pure' $2s$ nor 'pure' $2p$ AOs. The name used to describe such a combination of types in various fields is *hybrid* and this terminology is, in fact, used in this context, so:

Hybrid Atomic Orbitals (HAOs) (generically given the symbol ϕ)[4] are orbitals which are linear combinations of two (or more) 'ordinary' AOs of different types,

where, in this context the word 'type' means s, p, d, f etc. It is clear from the emphasis which has been placed on the polarisation of atomic charge distributions as part of the process of bond formation that these HAOs will

---

[3] It is atomic because both the $2s$ and $2p_z$ of which it is composed are *atomic* orbitals.
[4] Greek letter ϕ, pronounced 'fie'.

play an important part in any theory of the chemical bond.[5] Of course, as in the case of the hydrogen molecule, it may happen that, in a particular set of circumstances, no appreciable polarisation occurs and ordinary AOs may be retained to describe molecule formation.

3.2.2 *Molecular orbitals*

With the results of the last section we may now ask:

> What will an electron in the Li–H bond experience if it is shared between the two nuclei and is attracted to both?

There are obviously (at least) three interesting cases:

(1) When the electron is close to the lithium core or on the side of the lithium nucleus away from the hydrogen atom.
(2) Conversely, the electron is close to the hydrogen nucleus or on the side of the hydrogen atom remote from the lithium core.
(3) More interestingly, the electron is in the 'bond region', between the lithium core and the hydrogen nucleus, say half-way between the two.

The first two cases are the easiest to deal with. If an electron is close to one of the atomic cores, its motion and its distribution will be dominated by that core in much the same way that one of the moons of Saturn is dominated by the attraction of Saturn. The electron's motions and distribution will, of course, be influenced by any other atomic cores, but this will be a rather small effect compared to the 'local' atomic core.[6]

> So, *close to each atomic core the electron's distribution will be very similar to that associated with the (usually polarised) atomic orbital of that atom.*

Clearly, in the bond region *between* the atomic cores the electron 'feels' attractions to both atomic cores of comparable magnitudes, and so its motions and distribution must be different from those of either separate atom. We might say that the distribution or the orbital is characterised by its relationship to the AOs of *both* the separate atoms. What is needed

---

[5]See Appendix D for a clarification of the dispute about the use of HAOs.
[6]Remember, like gravitation, electrostatic interactions depend on the *square* of the distance between the particles.

is an orbital which is very similar to the (polarised) (H)AO[7] of each atom in regions of space close to that atom, but is different from either (H)AO in the bond (internuclear) region of space. But it is not *just* to be different, the required orbital must describe the actual electron distribution found in the real molecule tolerably well. Colloquially, something which looks like each AO close to each atom *and* which describes the bonding electronic structure well.

What are the properties of the electron distribution in the bond region expected to be? Consider the forces acting amongst the constituent (charged) particles of a diatomic molecule:

- The atomic core of each atom carries a net positive charge; for lithium it is the sum of the nuclear charge (atomic number: $+3$) and the charges of the inner shell electrons ($1s^2$: -2), overall $+1$; for hydrogen it is just the nuclear charge of $+1$. These like charges *repel* each other, giving a strong force *against* chemical bonding.
- The electrons in the bond each carry a negative charge and are attracted to *both* positively-charged nuclei, giving a strong force *towards* bond formation.
- The negatively-charged electrons repel each other, which gives a rather weakening effect on the bond formation.[8]

The electron distribution which would naturally seem to lead to maximum amount of the second of these two effects is the one where *both* electrons are as close to *both* nuclei as possible to offset the strong atomic core repulsion effects.

> In short, the situation one might expect is that electron distribution in the bond region be *larger* than the sum of the two separate atomic electron distributions in that region of space, so that the attraction between the nuclei and electrons overcomes the repulsions between the atomic cores and the smaller repulsions between the electrons.

Looking at the electron distributions of the sum of the separate atoms and the electron distribution in the molecule confirms this intuition. But

---

[7]HAOs are still *atomic* orbitals (AOs), of course, because they are composed of atomic orbitals from the same atom. In this section the two terms are used essentially interchangeably.

[8]The electrons are very 'mobile' and can avoid each other while still remaining in the bond region.

how is this to be represented? There is an obvious way to combine the two physical effects:

(1) The electron distribution of the bond near each atomic core and not in a bond region must be similar to that of the separate atom (polarised by the presence of any nearby atoms).
(2) The electron distribution of the bond between the two bonded atoms is expected to be greater[9] in the bond region than the sum of the two separate atomic distributions.

Clearly the second of these requirements means that we cannot get an adequate description of the electron distribution in the bond by simply adding the atomic electron distributions. But, equally clearly, we must add together *something* about the two electron distributions in order to satisfy the first requirement.[10]

> What has been found to work very well in practice in this situation is to describe the distribution of the two electrons forming the bond by an *orbital* which is the *sum* of the two (polarised) atomic orbitals, which (when squared) describes the electron distribution of the separate atoms. Technically, since chemical bonds do not usually share electrons equally between bonded atoms, we use a *weighted* sum — *linear combination of* — the (polarised) atomic orbitals of *both* atoms involved in the bond.

Let's say what this means and try to think why it should work. As usual, we start with the lithium hydride molecule, the simplest non-trivial case:

• Let ϕ_{Li} be the polarised valence-shell HAO of the lithium and ϕ_H the corresponding AO of the hydrogen atom. The electron distribution of the valence-shell of the lithium atom is then P_{Li}, say, given by:

$$P_{Li}(x, y, z) = [\phi_{Li}(x, y, z)]^2$$

and, using a similar notation, that of the hydrogen atom is

$$P_H(x, y, z) = [\phi_H(x, y, z)]^2.$$

---

[9] Or, at the very least, *different* from the sum of the atomic distributions!

[10] It is not possible, for example, to *multiply* properties of the distributions together, since the *dimensions* of the product would not match.

- Then we try an orbital ψ_{LiH} of the form:

$$\psi_{LiH} = c_{Li}\phi_{Li} + c_H\phi_H, \tag{3.1}$$

where c_{Li} and c_H are simply coefficients (numbers) chosen so that ψ is the best possible orbital for the bond.[11]

It turns out that this is all we need to obtain a good qualitative and quantitative description of the distribution of electrons in molecules and a satisfactory explanation of chemical bonds.

Perhaps what needs a little explanation is *why* this provides a description of the electron distribution in the bonding region. If the orbital for the bonding electrons is given by equation (3.1), then the electron distribution due to the electrons in this orbital is P_{LiH} (say), given by:

$$\begin{aligned} P_{LiH}(x, y, z) &= [\psi_{LiH}(x, y, z)]^2 \tag{3.2}\\ &= [c_{Li}\phi_{Li}(x, y, z) + c_H\phi_H(x, y, z)]^2 \\ &= [c_{Li}\phi_{Li}(x, y, z)]^2 + [c_H\phi_H(x, y, z)]^2 \\ &\quad + 2[c_{Li}\phi_{Li}(x, y, z)] \times [c_H\phi_H(x, y, z)]. \end{aligned}$$

If we suppress all the clutter due to writing out the dependence of the orbitals on (x, y, z), the above becomes more simply:

$$P_{LiH} = (c_{Li}^2) \times (\phi_{Li}^2) + (c_H^2) \times (\phi_H^2) + (2c_{Li}c_H) \times (\phi_{Li}\phi_H), \tag{3.3}$$

which may be translated into English as:

Distribution in the bond = (A number) × (Distribution in the Li atom)

+ (A number) × (Distribution in the H atom)

+ (A number) × (An additional term),

where the 'An additional term' is the *product* of the two separate-atom orbitals.

All of which says nothing more startling than the very familiar fact that

$$(x + y)^2 = x^2 + y^2 + 2xy \neq x^2 + y^2.$$

---

[11]This choice is not trivial, it requires some considerable computation, but that is not our concern here; we are just getting to grips with the *idea*.

So the explanation is:

> When we square the sum of two orbitals to obtain the
> electron distribution in the bond we obtain a distribution
> *different* from the sum of the two separate atomic distribu-
> tions, and this difference is greatest where the *product* of the
> two atomic orbitals is the greatest.

But where *is* this product at its greatest? A glance at some diagrams
showing the spatial forms of the polarised atomic orbitals answers this
question easily: it is in the internuclear region, and so it is this *overlap
region* which enables the sum of atomic orbitals to describe the electron
distribution in bonds adequately.

3.2.2.1 *Quick summary*

Before working out the details and consequences of these conclusions, a
quick summary may be in order. What is very obvious is that the description of the distributions of electrons in molecules is going to involve the
use of *atomic orbitals*. They are going to be the building blocks which are
used throughout the theory because:

- The atomic cores are to be described by the AOs of each atom *unchanged*:
  ordinary unpolarised AOs.
- The chemical bonds will be described by linear combinations of the
  'valence-shell' HAOs of the atoms of which the molecule is composed.

What remains is still considerable:

- How do we deal with 'lone pairs' of electrons?
- How, *in practice*, not in principle, do we use AOs to describe chemical
  bonding? In particular, what about molecules containing more than two
  atoms?

Also, on a note of detail. Let's not forget that equation (3.1) is written in
extremely compact notation. The atomic orbitals ϕ_A and ϕ_B are *functions
of position* in ordinary (3D) space and therefore so is the function ψ on
the right-hand side of equation (3.1). So, the equation should be written
in full as:

$$\psi_{LiH}(x, y, z) = c_{Li}\phi_{Li}(x, y, z) + c_H\phi_H(x, y, z) \tag{3.4}$$

if we use the conventional Cartesian co-ordinates to describe 3D space.

What must be emphasised is the *physical background* of the method we have described:

- The basic drive behind any explanation of the electronic structure of molecules is the fact that these systems consist of charged particles in mutual interaction governed by Schrödinger's mechanics (quantum theory).
- The use of atomic orbitals during the construction of the explanation is a matter of convenience, not principle; there are very good reasons why we can use these atomic orbitals as building blocks in the construction of the orbitals involved in a theory of valence.
- The interactions within molecules are between *charged particles*, not between atomic orbitals:

Electrons and Nuclei are physical objects, existing 'out there' in the real world. They can interact via their electrostatic fields.

Orbitals and their Squares are mathematical objects, existing in our minds, on paper or in our computers.[12] They cannot interact with each other.

However, notwithstanding these considerations, atomic orbitals are going to be our basic *mathematical* building blocks because:

- Colloquially, 'molecules are made of atoms'. So, in any molecular environment, there will be regions of space where valence electrons are close to atomic cores and whose motions and distributions are similar to those of the separate atoms; that is there will always be regions in molecules where electron distributions are very closely approximated by (squares of) (H)AOs.
- In bond regions, between atomic cores, the electron distributions will be different, but it proves possible to obtain a good approximation to the distributions in these bonding regions by using 'adjusted' (H)AOs.
- Here is the real bonus: we can do this adjustment by correcting the (H)AO on one atom by the addition of part(s) of the (H)AOs on other atoms and *vice versa*.
- In other words, by forming various combinations of (H)AOs, we can always obtain a good description of the electron distribution in all regions of the space occupied by a molecule's electron distribution.

---

[12]See Appendix C for details.

All this can be simply summarised by saying:

> Anywhere in and around a molecule — a combination of atoms — the electron distributions cannot be too different from (squares of) linear combinations of atomic orbitals.

3.3 Assignment for Chapter 3

This chapter is the basis of everything which follows about the electronic structure of molecules, so it is worth while going through the whole thing with your tutors *and* with your colleagues. In particular:

(1) Are you happy about making the transition from mathematical expressions expressed in words to their expression in symbols? Look at the equation for a polarised atomic orbital on page 62 and the opposite example in the expression for the electron distribution in a chemical bond on page 67. Do these make sense to you? Could you have done this yourself?

(2) Make sure that you appreciate *why* we use (H)AOs as our basic building blocks for the description of the electronic structure of molecules and, equally important, *why it works*!

(3) Can you think of any examples where this method might not work? Ask your tutor to explain.

Appendix D

Is Hybridisation a Real Process?

Ever since the beginning of quantum chemistry there have been disputes about hybridisation. On the one hand it has been said that the hybridisation (combination) of AOs is an essential process to understand why, for example, carbon usually has a valence of four and not two. On the other hand hybridisation has been seen as an arbitrary choice; the same result could have been obtained by using simple AOs.

There have always been disputes about the rôle of hybridisation, the step in the description of a chemical bond symbolised by:

$$AOs \longrightarrow HAOs,$$

the formation of a set of HAOs as linear combinations of AOs. There are two entrenched positions on the matter according to the answer given to the question:

Is hybridisation a physical (real) process or is it just an arbitrary way of describing the distribution of some of the electrons in a molecule?

This question cannot be given a satisfactory answer if posed in this way because:

(1) AOs and HAOs are both mathematical functions of ordinary (3D) space, not physical (real) objects.
(2) Any question about the existence of a real process can only be answered by considering the physical interpretation of the mathematics used in the description of the alleged process.

The physical interpretation of both AOs and HAOs is via their squares as probability distributions. In particular:

(1) The squares of AOs are the distributions of electrons in isolated atoms.
(2) The squares of HAOs are the distributions of electrons in atoms in the presence of a charge.[1]

Because AOs are used as basic building blocks with which to describe the electron distributions in molecules the obvious choice to model or *describe* the polarisation effects involved in bond formation is, therefore, HAOs. So, the answer to the original questions is:

> Hybridisation is not a physical (real) process, but the polarisation of an atomic charge density, which hybridisation is used to describe, certainly is.

It is not compulsory to use AOs or HAOs to describe the electron distributions in molecules — other mathematical functions have been used, even pure numerical solutions have been investigated — but it is very sensible and, above all, *interpretable*. The use of AOs and HAOs turns a theory of the electron distributions in molecules into a theory of *valence*.

---

[1]Strictly, the zeroth-order functions in the sense of perturbation theory.

Chapter 4

The Pauli Principle and Orbitals

The effect of the Pauli principle, which has been given in its simplest form on page 41:

> *When a many-electron system is described by electrons occupying orbitals, no more than two electrons may occupy any one orbital,*

is important throughout the theory of molecular electronic structure. The Pauli principle is a decisive factor in the theory of the electronic structure of molecules and needs more careful consideration than this simple statement. What needs to be done is to explain how and why the Pauli principle causes the saturation of bonding: why the valence of an atom is not the same as its atomic number, i.e. why the valence of carbon is generally four, not six, and of helium is zero, not two.

Contents

4.1   A Difficulty with Helium

We have seen that in the theory of the simplest stable molecule, H_2, two electrons occupying a single molecular orbital describe the electronic structure of the molecule, which is nothing more than the electronic structure of a single bond. When we go on to apply this same method to the interaction of two helium atoms we might expect that the four electrons in the system simply occupy *two* molecular orbitals of a similar type, bound by those four electrons giving a stable molecule more tightly bound than the hydrogen molecule. In fact, of course, we know that atoms of helium do not combine together at all and helium exists as a monatomic gas. Let's look at this more carefully.

There is nothing wrong with the first step. We can describe the distribution of two of the four electrons by saying that they occupy an MO qualitatively similar to the one in H_2. The Pauli principle prevents us from allowing any more electrons to occupy that MO. But, why not use an MO which is not the *same* as the first one, merely qualitatively *similar*? That is, why not use another MO which is cylindrically symmetrical around the internuclear axis and concentrating the electrons between the two nuclei? This could be different by being more or less extended in space, for example.

If the first MO is[1] ψ, then the description of the two electrons it contains is just the product of two copies of this function, one depending on the co-ordinates of electron 1 ($\psi(x_1, y_1, z_1)$) and the other involving the co-ordinates of the other electron 2 ($\psi(x_2, y_2, z_2)$):

$$\psi(x_1, y_1, z_1)\psi(x_2, y_2, z_2), \tag{4.1}$$

where (x_1, y_1, z_1) are the co-ordinates of one of the two electrons (electron 1, say) and (x_2, y_2, z_2) are those for the other electron (electron 2, say).

It is a *product* of these two functions because of the probability interpretation of orbitals. Remember that:

- The probability that an electron be in the region of space around the point (x, y, z) is the square of an orbital; so

---

[1] Remember, an *orbital* is a function of 3D space, so to emphasise this it should really be written in 'functional notation'. That is, we should write ψ as a function of three co-ordinates in much the same way as we write the sine function as a function of just one co-ordinate: $\sin(x)$. So, a fuller notation would be (if the three co-ordinates are to be the Cartesian ones, x, y and z) as $\psi(x, y, z)$.

- The probability that electron 1 occupying orbital ψ is near a particular point — co-ordinates (x_1, y_1, z_1) — is $|\psi(x_1, y_1, z_1)|^2$.
- The probability that electron 2 occupying orbital ψ is near a particular point — co-ordinates (x_2, y_2, z_2) — is $|\psi(x_2, y_2, z_2)|^2$.
- Therefore, the probability that electron 2 occupying orbital ψ is near a particular point — co-ordinates (x_1, y_1, z_1) — and *simultaneously* electron 2 occupying orbital ψ is near another point — co-ordinates (x_2, y_2, z_2) — is $|\psi(x_1, y_1, z_1)|^2 \times |\psi(x_1, y_1, z_1)|^2$ because probabilities are *multiplicative* in this situation.[2] But this simultaneous probability is just the square of the product of two copies of ψ, one for each electron.

So the quantum description of these two electrons occupying the one orbital is given by the expression (4.1):

$$\psi(x_1, y_1, z_1) \times \psi(x_2, y_2, z_2).$$

Now, suppose we have a second MO, χ, which contains the other two electrons (electrons 3 and 4, say, with co-ordinates x_3, y_3, z_3 and x_4, y_4, z_4). Then, by the same argument, these two electrons are described by a product completely analogous to the above, i.e.

$$\chi(x_3, y_3, z_3)\chi(x_4, y_4, z_4), \tag{4.2}$$

where the Greek letter χ (pronounced 'chi') has been used in place of the the ψ for the other MO. The key thing is that we are assuming *both* ψ and χ are similar in general qualitative form but not exactly the same, so that the way the Pauli principle has been expressed so far is not violated. There are just two electrons in ψ and just two in the other orbital χ. To investigate the consequences of this model of the helium diatomic molecule, let's write the difference between these two functions as δ (pronounced 'delta'), given by:

$$\delta = \chi - \psi$$

$$\text{i.e. } \chi(x, y, z) = \psi(x, y, z) + \delta(x, y, z), \tag{4.3}$$

where, again to remind ourselves that ψ, χ and δ are *functions* of 3D space, the explicit Cartesian co-ordinates are given in the second line.

---

[2]The particles are assumed to have *independent* distributions.

Now, if we substitute this expression for χ into equation (4.2), some simple algebra shows:[3]

$$\begin{aligned}
\chi(x_3, y_3, z_3)\chi(x_4, y_4, z_4) &= \psi(x_3, y_3, z_3)\psi(x_4, y_4, z_4) \\
&+ \delta(x_3, y_3, z_3)\delta(x_4, y_4, z_4) \\
&+ \psi(x_3, y_3, z_3)\delta(x_4, y_4, z_4) \\
&+ \delta(x_3, y_3, z_3)\psi(x_4, y_4, z_4).
\end{aligned} \qquad (4.4)$$

But we have met the first product term on the left-hand side of this expression. It describes two electrons in the orbital ψ; here, it is clearly interpretable as the MO ψ, containing electrons 3 and 4. This means that:

- ψ is occupied by electrons 1 and 2, as we have assumed in the first place: $\psi(x_1, y_1, z_1)$ means electron 1 occupying ψ, $\psi(x_2, y_2, z_2)$ means electron 2 occupying ψ *and*
- ψ is occupied by (at least) part of electrons 3 and 4 as well, because the above equation contains $\psi(x_3, y_3, z_3)$ and $\psi(x_4, y_4, z_4)$.

Looking in detail at what the expansion of $\chi(x_3, y_3, z_3)\chi(x_4, y_4, z_4)$ means, it is clear that, if the second MO χ contains *any part* of the other (doubly-occupied) MO, then occupying it violates the Pauli principle. Because, if the new orbital (χ) is to have *any contribution* from the existing occupied one (ψ), an equation like (4.3) ensures that, if it is occupied, it must cause more than double occupancy of ψ. The expansion of $\chi(x_3, y_3, z_3)\chi(x_4, y_4, z_4)$ (equation (4.4)) will contain a term in which electrons 3 and 4 occupy ψ.

We are then left with the problem:

> How do we ensure that this condition is satisfied? How can we be sure that each MO contains *no part* of another, 'earlier', doubly-occupied MO?

This calls for some serious explorations.

---

[3]Made to look a lot more complicated than it actually is by the necessity of including the co-ordinates of the four electrons. If it helps, just replace the electron co-ordinates by ordinary English. Replace $\psi(x_3, y_3, z_3)$ with $\psi(electron\ 3)$ and so on.

4.2 When are Orbitals Mutually Exclusive?

Remember we are dealing with *functions* of space, so some investigation is necessary to find out just what is meant by the idea that two functions can be 'mutually exclusive'; that each of the two functions contains 'no part of' the other. This is not quite so tricky as it first seems. We can look at some familiar functions to see what this sense of 'mutual exclusion' actually means. A simple example of functions of one variable will point us in the right direction.

The simplest trigonometric functions sine and cosine — $\sin(x)$ and $\cos(x)$ — have well-known expansions in terms of powers of x, which enable them to be calculated to any required accuracy:

$$\sin(x) = x - \frac{x^3}{6} + \frac{x^5}{120} - \frac{x^7}{5040} + \cdots \tag{4.5}$$

$$\cos(x) = 1 - \frac{x^2}{2} + \frac{x^4}{24} - \frac{x^6}{720} + \cdots \tag{4.6}$$

We may write each of these power-of-x terms with its denominator as a separate function, say

$$f_i(x) = \frac{x^i}{i(i-1)(i-2)\ldots 1} = \frac{x^i}{i!},$$

where $i!$ is just short-hand for $1 \times 2 \ldots \times i$ and $0! = 1$. The equation may be written in terms of these f_i as:

$$\sin(x) = f_1(x) - f_3(x) + f_5(x) - f_7(x) + \cdots \tag{4.7}$$

$$\cos(x) = 1 - f_2(x) + f_4(x) - f_6(x) + \cdots \tag{4.8}$$

Now, written like this it is obvious that, in any sensible use of the term 'part of', we can say that for example:

- The function $f_2(x)$ is *not* part of $\sin(x)$ and,
- In a similar way, the function $f_5(x)$ is not part of $\cos(x)$.

This is 'obvious' because $f_2(x)$ *does not occur* in the expansion of $\sin(x)$, nor does $f_5(x)$ occur in the expansion of $\cos(x)$. It is only necessary to *look* at the expansions in equations (4.7) and (4.8) to see this.

How can this be generalised? We will not often have such convenient expansions of our *orbitals* and, in any case, can hardly be examining them in great detail to find out if they are mutually exclusive or not. Is there a convenient *general* method we can deduce from these simple examples?

Yes, there is, and it involves a quantity which involves the whole of the functions. We may as well complete the job by continuing with the example of $\cos(x)$ and $\sin(x)$.

In the first of the two cases above it was asserted that simple inspection is enough to see that $f_2(x)$ is not part of $\sin(x)$. If we look at the product of these two functions:

$$f_2(x)\sin(x) = \frac{x^2}{2}\sin(x) \tag{4.9}$$

graphically, we can see immediately that this product has equal negative and positive contributions (say between $-\pi$ and $+\pi$).

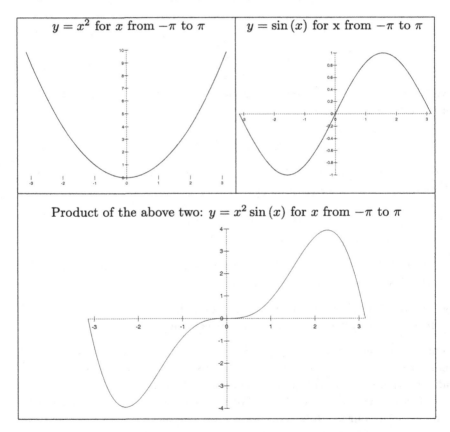

This means that the *integral* of this product *must be zero*, because the positive and negative regions of the product function (the parts above and below the x axis[4]) *exactly cancel*. This cancellation, the exact vanishing of the integral, is a property of the fact that one of the functions in the product $(f_2(x) = x^2/2)$ is *always positive*, while the other factor in the product $(\sin(x))$ *changes sign* at $x = 0$. So the total product *must* integrate to zero.

The second example shows the same effect: one factor in the product — $f_5(x)$ — changes sign at $x = 0$, while the other — $\cos(x)$ — is unchanged in sign at $x = 0$ so the product integrates to zero. This proves to be the general result:

> If the product of two functions integrates to *exactly* zero then these two functions are said to be *orthogonal*, which means that the functions do not contribute to each other at all; there is no linear expansion of one of the functions which contains any of the other function.

This result is usually interpreted by saying that the two functions do not *overlap*, in spite of the fact that they obviously *do* overlap, as can be clearly seen from their graphs. What is meant by 'overlap' here is 'overlap *integral*'.

> The overlap integral between two functions is just defined to be the *definite*[5] integral of the product of the two functions. So, if we have two functions, say $f(x)$ and $g(x)$, where x ranges from a to b, then the overlap integral of f and g is defined by

$$S = \int_a^b f(x)g(x)dx. \qquad (4.10)$$

Typically, if x is a Cartesian co-ordinate used for one dimension of space, $a = -\infty, b = +\infty$.

4.3 Does this Work for AOs?

We are already familiar with the *aufbau* principle for using the AO energy-level diagram, which involves 'at most two electrons per AO'.

---

[4]Remember the original basic definition of an integral. It is the area between the curve and the x-axis *with due regard for sign*.

[5]A definite integral is a number.

We have introduced these rules without once mentioning the question of the overlapping, mutual exclusion or orthogonality of the atomic orbitals. How have we got away with it without violating the Pauli principle, when we did not even *know* about the necessity of using AOs which do not overlap? Perhaps the simplest illustration is the valence shell of the carbon atom $(2s^2 2p^2)$: no attempt has been made to make sure that the $2s$ and $2p$ orbitals are non-overlapping (i.e., are orthogonal). If they are not, then we are in serious trouble, since the $2s$ AO — the lowest-energy one — will contain more than two electrons in violation of the Pauli principle. It must be said in advance of any thought about this problem that what we have done must be right, because it works so well in explaining the structure of the periodic table.

There is one complication here: the AOs are functions of 3D space so, we can not visualise them as easily as $\sin(x)$ and $\cos(x)$; we shall have to use the contouring method met in Appendix A.

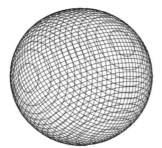
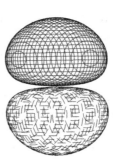

All-positive $2s$ AO:	$2p$ AO:
no sign-change on	sign-change on
horizontal reflection	horizontal reflection

If we look at contours of the two AOs we can see at once why these functions are orthogonal:

- The $2s$ function is *spherically symmetrical*; it takes the same value at a given distance from the nucleus in all directions.
- Each of the $2p$ functions is only *cylindrically symmetrical*; taking the $2p_z$ AO for concreteness, it *changes sign* on going through the origin along the z-axis, which is vertical in the above diagram.

So the *product* of these two functions has two 'lobes' of exactly the same size but of opposite sign, and therefore this product must integrate to zero in exactly the same way as the two products we have considered earlier ($x^2 \sin(x)$ and $x^5 \cos(x)$).

It is easy to see that this result is completely general:

> Atomic orbitals of different 'types', s-type, p-type, d-type etc., are orthogonal, and so the Pauli principle in its simplest possible form applies to their use.

This is easy to see because the AOs of different types have different *symmetries*. They behave differently when, for example, reflected in planes through the nucleus: they have more lobes or lobes of different signs.

This, however, cannot be the end of the matter because, in the case of the carbon atom, there are *two* AOs of the same type — $1s$ and $2s$ — which are both doubly-occupied and are therefore not covered by the above result. In fact, these AOs are orthogonal, but the reason is not quite so obvious; it is not just a simple question of symmetry because the two s-type AOs have the same symmetry. This time, to make the orthogonality of the two AOs be seen we must look at the way they each behave *as a function of distance, r, from the nucleus.*[6]

The 1s and 2s radial AO as a function of distance from the nucleus

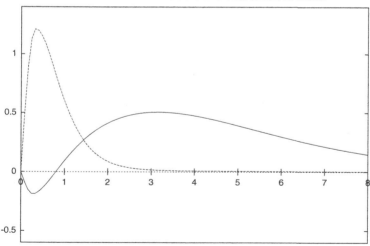

---

[6]These two AOs are actually plotted as $r \times 1s$ and $r \times 2s$ respectively for the mathematical reasons which define overlap integrals in the radial co-ordinate system.

It is clear that the $1s$ function is always positive (it lies above the axis for all values of r). But the $2s$ AO changes sign: the inner part is negative while the outer part is positive.[7] Thus, the product $1s(x, y, z) \times 2s(x, y, z)$ changes sign where the $2s$ AO changes sign. What is not obvious but is nevertheless true, is that the inner (negative) parts of the product exactly cancel the outer (positive) parts of the product, so that the whole thing integrates to zero. Again, it is not as obvious as in the AOs of different types, but it is the case that all AOs[8] of the *same* type are also orthogonal.

So, after this lengthy digression we can see that what we did in all innocence is all right after all.

> The use of the simple Pauli principle in describing the structure of many-electron atoms is correct because the AOs are all mutually orthogonal.

Now we must get back to the problem in hand: what is the electronic structure of the (putative) He_2 molecule?

4.4   The Helium Molecule — Again

In view of the analysis in Sections 4.2 and 4.3, if we are to get to the bottom of the problem of the matter of the electronic structure of the He_2 molecule, we must find a molecular orbital (MO) which is orthogonal to the one formed from a positive linear combination of the $1s$ AOs of the constituent He atoms, i.e.

$$\psi(x, y, z) = c_1 1s_1(x, y, z) + c_2 1s_2(x, y, z). \tag{4.11}$$

In order to be orthogonal it is sufficient that the MO we seek must be of a different *symmetry* to this MO.

But what *is* the symmetry of the MO given by equation (4.11)? Just like the H_2 molecule, the fact that He_2 is a *homonuclear* diatomic molecule means, along with the other requirements of a suitable MO, that the MO must be made up of *equal* contributions from the two He

---

[7]The familiar spherical picture of any s AO is due to the fact that we can only see an *outer* contour in these representations; the negative part(s) are *inside* the drawn contour.

[8]Note the fact that we are talking about atomic orbitals, not all orbitals.

$1s$ AOs: i.e. $c_1 = c_2$, so the overall MO (4.11) has the following symmetry characteristics:

- Viewed 'from the side' the MO is unchanged by reflection in a mirror at the bond midpoint (perpendicular to the bond).
- Again, viewed from the side, the MO is unchanged by reflection in a mirror containing the bond.
- Viewed from the end the MO looks like a $1s$ AO; it can be rotated around the bond without changing.

So if we want an MO which is orthogonal to this one, we can obtain one which differs by the fact that it *changes sign* by one or more of the above reflections and rotations. However, remember what we are trying to do here: we are looking for an MO which will generate an electron distribution that will assist in bonding the two atoms in the molecule; we are not just fulfilling a symmetry recipe. There are plenty of possibilities; here are a couple:

(1) An MO which changes sign in the perpendicular mirror at the bond midpoint

$$\psi_1(x, y, z) = 1s_1(x, y, z) - 1s_2(x, y, z) \qquad (4.12)$$

using the He $1s$ AOs but with opposite signs (neglecting the coefficients, for simplicity).

(2) An MO which changes sign in the mirror containing the bond

$$\psi_2(x, y, z) = 2p_{x1}(x, y, z) + 2p_{x2}(x, y, z), \qquad (4.13)$$

where the two $2p$ AOs of the excited states of the He atom have been called into service. The z-axis is perpendicular to the bond.

It is worth while reviewing the requirement given in Section 3.2.2:

(1) The electron distribution of the bond near each atomic core, and not in a bond region, must be similar to that of the separate atom (polarised by the presence of any nearby atoms).
(2) The electron distribution of the bond between the two bonded atoms is expected to be greater in the bond region than the sum of the two separate atomic distributions.

The first MO, equation (4.12), is the opposite of what is required; it has a *smaller* electron distribution between the nuclei than the separate atoms.

The second, equation (4.13), has a larger electron distribution between the two atoms than the separate $2p$ AOs, but smaller than the two $1s$ AOs. Neither of these MOs look like suitable candidates for a second MO with which to bind the two helium atoms which, in their ground states, have an occupied $1s$ AO. When detailed calculations are performed, it is found that there is no MO for the putative He_2 molecule which is suitable.

> The lowest-energy MOs which can be found and whose occupation satisfies the Pauli principle for two helium atoms on close approach do not provide enough binding energy to overcome the repulsions between the nuclei. There is only one MO of energy lower than the energies of the separate-atom $1s$ AO; all others are of higher energy.

Therefore helium exists, not as diatomic molecules but as separate atoms.

Of course, if one could actually *hold*[9] two helium atoms at a typical molecular bond distance and prevent them from separating, then the electrons would *have to* take up distributions — i.e. occupy orbitals — with higher energies than those of the separate atoms, and one can compute these rather artificial cases. The lowest two MOs in this case are the familiar one analogous to that occupied in H_2, and the one described qualitatively by equation (4.12). In this situation there would be very large forces exerted to force the atoms apart. For this reason, the orbital (4.12) is sometimes called, rather misleadingly, an 'anti-bonding' MO. It is misleading because it is not *occupation* of this MO which forces the atoms apart, but the fact that it is never formed because it is of too high energy to be occupied, which prevents He_2 from forming at all.

Notice that it is the Pauli principle which is responsible for this outcome. There *is* a molecular orbital which is low enough in energy and of the right form to bind the molecule but, to achieve the binding effect, it would have to be occupied by all four electrons. This is explicitly forbidden by the Pauli principle. Notice how completely different this situation is from the one we are all familiar with at the level of larger, heavier 'macroscopic' objects. If we had the means to do it, there is no 'large-scale Pauli principle'

---

[9]With atomic tweezers, perhaps!

which would prevent us from putting any number of planets in the same orbit around the sun. But the Pauli principle says:

> Even though mechanics at the sub-atomic level — Schrödinger's mechanics — says that energies and associated orbitals (electron distributions) can be found which would allow helium to form a diatomic molecule, the Pauli principle forbids it.

This is to be a feature in the theory of the electronic structure of all molecules. It explains why chemical bonding *saturates*. Why an atom with, say, Z electrons does not, in general, form Z bonds with other atoms. Helium has two electrons. Chemical bonds are formed by sharing electrons but helium does not form a double bond. Manganese has 25 electrons, but manganese does not form 25 bonds; no one has seen MnH_{25}, for example. This is similar to the familiar *aufbau* recipe for describing the electronic structure of many-electron atoms. It is the Pauli principle which, by preventing AOs being occupied by more than two electrons, generates the repeating 'outer' electronic structures which are summarised in the periodic table.

4.5 The Rôle of Atomic Orbitals in Valence Theory

It has already been stressed[10] that orbitals of any kind — atomic (AOs), molecular (MOs) — do not exist as physical quantities 'out there' in the real world. They are functions; mathematical creations to help us to understand the energies and distributions of electrons in atomic or molecular environments. Like any other probability distributions — from the tossing of coins to the occurrence of road traffic accidents — they only exist in our minds, our computers or on paper.

But the theory of the distribution of electrons in molecules depends heavily on the use of AOs. The MOs which have been used to describe the electron distribution in an atomic core *and* in a 'bond pair' of electrons are actually constructed from (linear) combinations of atomic orbitals. Of course they are, because:

- Molecular orbitals are also *functions*; when squared they give the probability distribution of electrons in molecules.

---

[10]See Appendix C for a refresher on details.

- More importantly, combinations of AOs satisfy the conditions which an MO describing a localised electron-pair bond should satisfy, as outlined on page 65.

And that is all.

> Atomic orbitals are used to form molecular orbitals because they are a mathematically-useful and intuitively-attractive *tool* with which to short-cut the potentially difficult task of solving the equations of quantum theory for the distributions of electrons in molecules.

The simple addition of AOs to form MOs is appealing because it relates the electronic structures of molecules to those of the component atoms of the molecules. This technique then becomes a theory of *valence* — the way the structure of molecules may be related to the structure of atoms — rather than simply a theory of the electronic structure of molecules, which is an altogether less chemically appealing theory. It also make the *detailed calculation* of electronic structure possible.

It is clear that in the wider context of the electronic structure of matter, this technique might well cease to be useful or appropriate. Attempts to develop a theory of the electronic structure of (solid) metals, for example, do not get very far using AOs as building blocks for the 'valence' electrons, because the criteria we used for molecules (page 65) do not apply: the characteristic feature of the electronic structure of metals is that they *conduct electricity* (the electrons are not localised) while, by and large, molecules do not. This means that (localised) AO-based descriptions of the electronic structure are of very limited value.

The drawback to the use of AOs in valence theory is that the attractive simplicity of the method has been abused. AOs are sometimes thought of as existing 'out there' while being able to *interact* with each other and form, as well as approximations to the MOs in molecules, other, more mythical entities, as we shall see.

> The use of AOs to form MOs is a way of turning 'the quantum theory of the distribution of electrons in molecules' into a theory of *valence*. It is the detailed *technique* which is used to satisfy the general requirements which were introduced in Chapter 1. It provides a *chemical model* of the electronic structure of molecules which helps to break down the structure into meaningful and manageable substructures.

As we saw at the end of Chapter 3, electrons and nuclei exist in the real world and may interact with each other; orbitals do not and cannot.

4.6 Current Summary for LiH and 'He₂'

This chapter and the previous one will prove to be a basis for most of what is needed for a theory of the electron-pair bond in chemistry. There are some other matters which will require further analysis, in particular:

(1) How do we cope with the existence of *delocalised* electronic substructures in some molecules? The methods used so far cannot deal with the π-electron structures of conjugated polyenes and aromatics.

(2) What about valence-shell *lone pairs*? When the electron distribution is polarised by and shared with another atom, this *must* affect the distribution of any valence-shell electrons not involved in bonding;[11] how is this to be described?

However, in the central area of the theory of the localised electron-pair bond, we have enough material to proceed. But remember the problems we raised in Section 1.5 on page 16. The Pauli principle resolves only the second of those three:

- What about *polyatomic* molecules? How are atomic electron distributions changed and combined when an atom is bonded to more than one other?

- Why does bonding *saturate*? Why does, for example, a carbon atom overwhelmingly form four bonds and not three or six?

- How is the *shape* of polyatomic molecules determined? The simple diatomics we have looked at have no choice about their shape.

The first and third are still outstanding and we must at least begin to address them. The simplest of these is a consideration of (valence-shell) lone pairs: electrons in atomic orbitals which have similar energies to the electrons involved in bonding, but which themselves are not shared between two (or more) nuclei.

---

[11] Because all electrons repel one another.

4.7 Assignment for Chapter 4

This is the only case in quantum chemistry where we have simply to accept a law of nature which is not connected to our experience as students of science to provide some justification. *But* it is the only way to get our theories to agree with experiment. So, we need, at the very least, to become familiar with its application if we cannot understand its existence.

The importance of the Pauli principle in chemistry is that, among other things, it explains why valence *saturates*: why an atom (A, say) of atomic number Z does not form molecules like AH_Z for $Z > 3$.

Make sure that you appreciate these facts and discuss any difficulties. Problems to think about:

(1) Laws which have to be assumed 'on trust' can be justified by the fact that they must be used in order to explain the observed facts. One says that such laws are justified (or verified) *a posteriori*.[12] Can you give examples of other such laws in chemistry? Do you find this method of justifying laws 'in reverse' acceptable?

(1) The lithium atom has the electronic structure $1s^2 2s$ in which the $2s$ electron is mainly 'outside' the spherically-symmetrical lithium core of two $1s$ electrons. Thus, the net field in which the $2s$ electron moves is a pull by the nucleus of charge $+3$ and a repulsion by two inner electrons each of charge -1. This is very similar to the net field experienced by an electron in the hydrogen atom (net attractive charge $+1$); why is the electronic lowest valence AO for Li not a $1s$ AO as it is in the case of the H atom?

---

[12]Latin, meaning roughly, 'in arrears' or 'afterwards'.

Chapter 5

A Model Polyatomic: Methane

Rather than proceed to considerations of the bonding in other diatomic molecules which have more complex bonding patterns, this chapter looks at perhaps the most important generalisation of the ideas developed so far: the electronic structure and bonding in polyatomic molecules. The simplest of these is the 'fundamental' organic molecule methane, CH_4. The theory of the bonding in homonuclear and heteronuclear diatomics — which involve new types of bond — is deferred until Chapter 9.

Contents

5.1 The Methane Molecule: CH$_4$

There are three main things to be explained about the familiar hydride of carbon:

(1) Why it is CH$_4$ and not CH, CH$_2$ or CH$_3$, since the carbon atom contains an atomic core of two ($1s$) electrons and four valence electrons? Why are all four involved in bonding rather than one, two or three?
(2) What are the actual details of the electronic structure and bonding in the molecule?
(3) What shape is the molecule? It might be thought obvious 'by symmetry' that a molecule composed of four atoms joined to one central atom would be tetrahedral, since whether the hydrogen atoms repel or attract each other it is easy to see that four identical atoms equidistant from a given centre and free to move about *must* take up a tetrahedral arrangement.[1] But, as we shall see later, not all molecules AB$_4$ are tetrahedral, so this simple argument cannot be the whole story.

The first of these problems involves something quite simple and not essentially our concern here. The fact is that the molecules CH, CH$_2$ and CH$_3$ *do* have independent existence but they are extremely reactive; or, as one would say in a chemical context, they are very unstable. It is worthwhile reminding ourselves of what is meant by 'stable.' The word means different things in different contexts:

- If 'stable' is to mean 'existing on its own in the absence of any other molecules', i.e. stable with respect to falling apart into separate fragments or atoms, then CH, CH$_2$ and CH$_3$ are stable and so is CH$_4$. This is, perhaps, the physicist's or spectroscopist's idea of stability; these molecules would exist in isolation in outer space, for example. But we are only considering isolated molecules so any theory we develop should be capable of describing the stability of these molecules; but not yet!
- If, on the other hand, 'stable' is to mean 'reluctant to react with many other molecules', i.e. relatively unreactive, then CH, CH$_2$ and CH$_3$ are not stable. These molecules will react with each other and practically any other molecule that they come into contact with. This is the meaning of stability used by chemists in general.

---

[1] Convince yourself that this is true.

Thus, although any theory of the bonding in molecules should be able to cope with these very reactive molecules, it is probably best to put off the most general theory until we have dealt with some more familiar and chemically stable examples.

5.2 The Electronic Structure of Methane

The simplest point of view to take of the electronic structure and bonding in the methane molecule is that it is analogous to four lithium hydride molecules sharing a common, heavier atom. The conventional structural formula

supports this idea, with the central sphere representing the carbon atomic core and the H symbols being the hydrogen atomic cores — nuclei in this special case — and the four lines, as usual, representing bonding pairs of electrons. The details are fairly straightforward but, as we shall see, there are some complications.

The central idea is to use the method outlined for the hydrogen and lithium hydride molecules in Section 1.4 on page 61 and to use the concepts of hybrid atomic orbitals (HAOs) developed in Section 3.2.1 on page 8. We consider what happens when four H atoms approach a C atom:

(1) The charge distribution of the central carbon atom will be polarised by the approach of the hydrogen atoms, each H atom 'pulling' a portion of the carbon's four valence electrons along its line of approach in exactly the same way that the approaching hydrogen atom distorts the distribution of the single valence electron of lithium, as illustrated by the diagrams on page 11. That is, in the language of atomic orbitals, each H atom generates an HAO which is 'pointing' at itself.

(2) Again, as we have seen, there is a reciprocal action by the C atom on the electron distribution of the H atom, but this is minimal for the hydrogen atom and not readily seen on the contour diagrams.

(3) As the H atom approaches the electronic structure of each atom is disrupted and a new molecular distribution is formed, which can be

described by the formation of a molecular orbital (MO) by a combination of the two separate-atom HAOs, as described in detail in
Section 3.2.2 on page 63.

The result is an electron distribution in the molecule which is composed
of four MOs each occupied by two electrons — one from the carbon atom
and one of the four hydrogen atoms. Here are contour diagrams of the
separate-atom HAOs and the MO at the equilibrium bond distance of the
C and H atoms in methane.

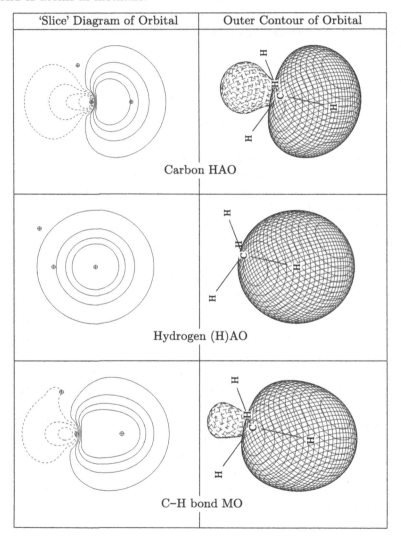

'Slice' Diagram of Orbital	Outer Contour of Orbital
Carbon HAO	
Hydrogen (H)AO	
C–H bond MO	

Because of the importance of the methane molecule as the 'basic' organic molecule, the HAOs and resulting bond MO have been given in the two possible representations.[2] Notice, as mentioned in Appendix A, the straightforward graph of value-of-orbital plotted against distance which would be very confusing even in this relatively simple case. The only really useful graph of this sort would, perhaps, be the value of the bonding MO along the C–H bond.

5.3 The Shape of the Methane Molecule

The electronic structure of the bonds in methane — the simplest organic molecule — has proved easy to describe with the concepts we have developed so far, and this method will be one of the most powerful tools in our approach to the structure of saturated molecules. But this has only given us the electronic structure of the bonds, the familiar lines in the structural formula representing a two-electron bond. 'What about the way these individual structures *interact* with each other?' After all, two bonds in close proximity (sharing a common end atom) are still composed of charged particles and these charged particles will interact both within a given bond and *between* the bonds. The theory so far has only dealt with the interaction within each bond. What are the forces acting between bonds? Of course, they have just the same origin as those acting within the bonds, electrostatic attractions and repulsions:

(1) There is a mutual attraction between nuclei within one bond and the electrons of the other bond.
(2) The pair of electrons in one bond will repel the pair in the other bond.
(3) The nuclei attached by one bond will repel the nuclei attached by another bond (except, of course, when two bonds share a common atom, like the C atom in CH_4).

Not all of these three effects operate in the same direction. In particular, it is easy to see that the effect of the first one is for the 'hydrogen ends' of each of the bonds to get as close as they can, i.e. this effect will try to fold up the molecule like a furled umbrella.

---

[2]Note that the 'slice' contour diagram does not show all five atoms in CH_4; of course not, because a (planar) slice cannot go through all five atoms.

The other two effects, however, do work in the same direction:

(1) The mutual repulsion between the bonding electron pairs means that the 'hydrogen ends' of each bond will be forced away from each other.
(2) Similarly, the repulsions amongst the hydrogen nuclei will force them apart.

If we make a very rough estimate of magnitude of each of these three effects by simply counting the number of interactions between any two bonds sharing a common atom, we can see that:

(1) Attraction between one H nucleus and two electrons is 'two terms', giving a total of four terms for this effect.
(2) Repulsion between two pairs of electrons is four terms.
(3) Repulsion between two H nuclei is one term.

The numbers are close but the indication is that the repulsion is likely to overcome the attractive terms by five to four. There is, therefore, a net repulsion between the 'bond pairs' sharing a common atom, i.e. two atoms joined by single chemical bonds to a common atom (a structure like A–B–C) will experience a net repulsion which will keep them as far away from each other as is allowed by the constraints of the bonding.

So, in this case, as we noted at the very outset, it is easy to see that the way in which four atoms joined to a common atom can satisfy the requirement of being as far away from each other as possible is for the four to take up a tetrahedral arrangement about the central atom.

It is worth noting that these considerations would lead to the molecule AB_2 being linear and AB_3 being a plane triangle. The fact that OH_2 (i.e. H_2O) is actually not linear and NH_3 is not planar obviously needs some attention, and this will be considered in the next chapter.

5.4   What About the Pauli Principle?

These explanations seem to hang together and give a satisfactory qualitative explanation of the bonding in the methane molecule. But we have not paid any serious attention to the all-pervasive Pauli principle. This constraint is at least partially satisfied by the fact that we have only allocated two electrons to each bond MO, but we have not considered whether or not these bond MOs are mutually exclusive, as we did for the theory of 'He_2'. That is, do the bond MOs *overlap*? It is obvious that they must overlap because the MOs are concentrated in a very small region of space, particularly close to the carbon atom. Here, for example, is a diagram of two of them superposed:

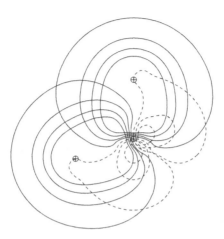

But, remember, it is a zero overlap *integral* which is involved in satisfying the condition of mutual exclusivity. There are positive and negative 'lobes' of each MO, so there is the possibility of cancellation but, as we have formed them, the four MOs do, in fact, have non-zero overlap integrals (all the same, of course, because of the high symmetry of the molecule).

It turns out to be possible to make any set of doubly-occupied MOs orthogonal by a mathematical device which does not affect the interpretation of those MOs. This technique was hinted at when it was shown, in Section 4.4 on page 82, that occupying two MOs composed of the two $1s$ AOs of the separate atoms of 'He$_2$' is entirely equivalent to double occupancy of the two $1s$ AOs themselves. A full justification of this idea — central to the description of the bonding in polyatomic molecules — is too mathematically complex for the time being.

> Unfortunately, like the Pauli principle itself, we shall have to just accept this as a fact, although — unlike the Pauli principle — it is a *theorem* which can be proved, not a law of nature. So, we say that we may ignore the effect of the (small) inter-MO overlaps if all the MOs are doubly-occupied.

5.4.1 Preliminary summary for methane

The only new ideas involved in extending the concepts of Chapters 3 and 4 are:

(1) A more careful consideration of the Pauli principle as it impacts on several doubly-occupied MOs.
(2) The use of straightforward ideas of electrostatic interaction to get to grips with an explanation of the shapes of molecules.

5.5 The Chemist's Description of Methane

Having gone to some trouble to set up a theory of the simplest organic molecule, it behoves us to check if it is acceptable to that notoriously fastidious person, the 'working chemist at the bench'. Is this the way that a chemist likes to describe the bonding in saturated organic molecules? The answer is, as one might expect, 'partly'.

In our description of the bonding in methane, stress has been placed on two central ideas:

(1) The breakdown of the total electronic structure into environment-insensitive substructures: bonds and atomic cores.
(2) The use of AOs in the scheme:

$$AOs \longrightarrow HAOs \longrightarrow MOs$$

to generate descriptions of the individual (localised) electron-pair bonds.

These points fit in very nicely with the experimental fact that, for example, the C–H single bond is 'transferable' from molecule to molecule in the sense that its length and bond energy are almost constant across a wide range of saturated molecules.

However, not content with a description of the static molecule, chemists are interested in explanations of the type of *reactions* which molecules undergo. Ever since the very simplest theories of valence, chemists have found it useful to systematise and rationalise the behaviour of molecules by invoking electronic structures and bonding schemes which differ from the accepted stable structures; structures which reflect the ease with which it is possible for a reacting molecule to be polarised by an approaching reactant. This is entirely rational. A central plank of our theory of valence is the idea that the electron distribution of atoms and molecules are polarised by each other's approach. So the polarisation of the electron distribution of a *whole molecule* with the formation of more bonds — i.e. a chemical reaction — is just another example of this.

This method involves choosing some 'chemically reasonable' structures for a given molecule which give an electron distribution different from — typically, but not always, more polar than — the accepted normal structure. There are some rules:

- The idea of environment-insensitive substructures is retained.
- The use of the scheme

$$AOs \longrightarrow HAOs \longrightarrow MOs$$

  to generate descriptions of the individual (localised) electron-pair bonds is retained.

- The structures should not be *too* fanciful; they have to be 'chemically reasonable', not something which can be found in the tails of comets or particle accelerators.

Unfortunately for us, methane is a poor choice, since it does not have much in the way of reactions; it burns, and that is about all. But we can still use the molecule as an illustration of the technique.

What are the chemically reasonable structures for a group of two atoms, A and B, say?

The one we have used	a covalent bond	A–B
One singly ionic structure	an 'ionic' bond	$A^+ B^-$
The other singly ionic structure	"	$A^- B^+$
Other doubly ionic structures e.g.	"	$A^{2+} B^{2-}$

The more highly-charged structures are judged less and less chemically reasonable.

In our case, we can invoke these choices *within* our valence substructures, i.e. within each bond:

The Covalent Bond	C–H
Polar Structure 1	$C^+ H^-$
Polar Structure 2	$C^- H^+$

Other, more highly charged structures are unlikely.

What remains is to use the scheme $AOs \longrightarrow HAOs \longrightarrow MOs$ to describe these ionic structures and to find a way to use them in the description of the methane molecule. What are the changes in the physical processes leading to the polar structures rather than to the covalent one? Take what is, perhaps, the more likely of the two, C^-H^+:

- The approach of the two atoms will still polarise their electron distributions, leading to the step AOs $\longrightarrow$ HAOs.
- The difference is, we assume that, instead of occupying an MO formed from the HAOs, the two electrons undergo a rather more drastic reorganisation and the electron from the hydrogen atom is transferred completely from its original H atom HAO to the HAO on the C atom. The resultant structure is then *two* electrons occupying the carbon HAO and *none* in the hydrogen HAO, giving a more detailed description of the electronic structure than simply saying C^-H^+.

Obviously the other ionic structure is described similarly, but with the final electron distribution being no electrons in the carbon HAO and two in the hydrogen HAO. One might imagine that the implied existence of the H^+ ion in the one case and the H^- in the other has some bearing on which of these two ionic structures is the more stable.

Clearly then, simply by using only part of the sequence to form MOs we can obtain descriptions of the other, less stable, electronic structures of the methane molecule and, by implication, of any other saturated molecule. Equally clearly, it is easy to think of other molecules for which the *ionic* structures of page 97 would be more appropriate than the covalent one: $NaC\ell$, for example.

5.5.1　*How to use these structures: the valence bond method*

We have now established the idea that it is possible, using the two HAOs involved in a localised single bond, to describe three alternative electronic structures for that bond. Each of these structures is associated with a 'chemical structure' of a familiar classical type. Since it is clear from experimental chemistry that some molecules will have the covalent structure as their preferred state, and some will be happiest with an ionic description, the question naturally arises:

> How can we use the *theory* of electronic structure to take care of all cases without appealing to experiment? After all, a theory is pretty poor if it cannot *predict* the structure.

What we need is a method of 'allowing' the theory to decide 'for itself' which is the most appropriate description of a particular electron-pair bond. It will come as no surprise to find out that, although there are some extreme cases, detailed calculations show that most bonds take up an electron distribution which is intermediate between the three cases. In short, we must use a description of a chemical bond — between A and B, say — which we might say is, colloquially speaking,

Chemical bond structure = Part of the covalent structure A–B

+ Part of the ionic structure A^+B^-

+ Part of the ionic structure A^-B^+

In fact, this recipe is exactly what *is* used when expressed in suitable mathematical notation. In a way that is completely analogous to:

• The formation of HAOs from AOs:

HAO on an atom = (a number) × (An AO on an atom)

+ (a number) × (Another AO on that atom) + ⋯

• and the formation of MOs from HAOs:

Molecular orbital = (a number) × (HAO of atom A)

+ (a number) × (HAO of atom B)

• we write:

Chemical bond structure = (a number) × covalent structure A–B

+ (a number) × ionic structure A^+B^-

+ (a number) × ionic structure A^-B^+

The difference is, of course, that the first two are linear combinations of *orbitals* of some kind, while the last is a linear combination of chemical structures.

When hybrid atomic orbitals (HAOs) were first introduced, the name 'hybrid' was justified by the fact that, typically, an HAO is a combination of two different *types* of AO and so is, in the usual sense of biology, indeed a hybrid. Now, perhaps confusingly, the same kind of justification applies to the linear combination of different *types* of chemical structure to describe a chemical bond. So a chemical structure described in this way is also called a hybrid because it is partly covalent and partly ionic.[3] This model is known as the *valence bond* (VB) method since it may invoke several different structures with different numbers or type of chemical bond contributing to the overall valence structure.

> There is an acute danger of confusion here, since, in the VB method, the simple chemical structures will usually be described by the use of hybrid *atomic orbitals* combined to form a chemical structure which is described as a hybrid of simpler *chemical structures*.

The above picture of the electronic structure of a bond can be put into more formal language by using a symbol for each of the structures and the 'mixing coefficients' denoted above by 'a number'. It is clear that the linear combination of structures looks very like the linear combination of HAOs. So, to distinguish between these two cases, we can use (Greek) capital (upper-case) letters for the (two-electron) structures while retaining lower-case letters for (one-electron) orbitals. In the above expression we use:

- Ψ_{VB} for the overall electronic structure of the bond;
- Ψ_{HL} for the covalent structure (**H**eitler and **L**ondon were the first to use this method) with coefficient C_{HL};
- Ψ_{+-} for one ionic structure with coefficient C_{+-}; and
- Ψ_{-+} for the other ionic structure with coefficient C_{-+},

so that we have in place of the above descriptive scheme:

$$\Psi_{VB} = C_{HL}\Psi_{HL} + C_{+-}\Psi_{+-} + C_{-+}\Psi_{-+}, \tag{5.1}$$

leaving the coefficients C_{HL}, C_{+-} and C_{-+} to be determined by detailed calculation. It is easy to guess the *relative* size of these coefficients in some familiar cases:

- A symmetrical single bond like the one in H_2 or a bond between atoms with similar electronegativities like the C–H bond in saturated hydrocarbons will be 'mostly covalent' so C_{HL} will be larger than either of the

---

[3] Many texts, particularly older ones, use the term 'resonance hybrid' due to some imagined analogies from the early days of quantum theory.

other two, C_{+-} and C_{-+}, which will be the same or similar in value:

$$C_{HL} > C_{+-} \approx C_{-+}. \tag{5.2}$$

In the case of the C–H bond in methane the values are:

$$C_{HL} = 0.75, \quad C_{+-} = 0.34, \quad C_{-+} = 0.57.$$

The squares of these numbers give the relative proportions of the three structures in the composition of the bond. When multiplied by 100 the percentages are:

$$\text{C–H } 56\%, \text{ C}^+\text{H}^- \text{ 11.5\%, and C}^-\text{H}^+ \text{ 32.5\%,}$$

temporarily using simply C–H for the HL homopolar structure. It is easy to show that the MO function (approximately equal contributions from each HAO in the AO) corresponds to *equal* contributions from all three VB structures.

• A purely ionic bond like the one in, say, NaCℓ will have a large C_{+-} (the principal structure being Na$^+$Cℓ^-) with much smaller C_{HL} (not much covalent contribution) and a *very* small C_{-+} (hardly any Na$^-$Cℓ^+):[4]

$$C_{+-} \gg C_{HL} \gg C_{-+}. \tag{5.3}$$

Thus the simple MO model of the two-electron bond may be improved by admitting a 'mixture' of the other possible — but less stable — electronic structures originally invoked to explain the reactivity of those bonds.

5.6   Summary for Methane

What we have learned in this chapter will prove to be the basis of the theory of all localised two-electron bonds wherever they occur and whatever the type of HAO used. In addition to the ideas developed from the study of the LiH molecule:

(1) The division of the electronic structure into bonds and atomic cores;
(2) The polarisation of the atomic electron distributions and its representation by the formation of HAOs and then MOs; the now

---

[4]The symbol $\gg$ means 'much greater than' and $\approx$ means 'approximately equal to'.

familiar sequence

$$AOs \longrightarrow HAOs \longrightarrow MOs,$$

we have found two new developments:

(3) The repulsions amongst the two-electron bonds determine the shape of the molecule.

(4) Considerations of electronic structures more general than the covalent bond introduce the idea of the mixing of three structures to generate a possibly more precise description of the bonding: the VB method.

What remains before we can be confident of describing molecules containing only localised two-electron bonds is:

• An account of the rôle of lone pairs of electrons.
• A theory of multiple bonds between atoms.

These questions will be tackled, as far as possible, separately.

5.7   Assignment for Chapter 5

(1) There are two approaches to the electronic structure of the methane molecule in this chapter; which do you prefer and why? There is also the possibility of serious confusion of the *nomenclature* involved. Make sure that you can distinguish between a hybrid atomic orbital and a hybrid structure in the VB method, since a hybrid structure will usually involve the use of hybrid atomic orbitals!

(2) The VB method is often used by experimental chemists as a way of describing reactions undergone by the various substructures of a molecule. For example if a molecule is involved in a reaction in which it has a hydrogen cation removed, one might say that, during the reaction, the structure which is *predominantly* C–H is changed to one which is *predominantly* $C^- H^+$, from which it is easy to remove a hydrogen atom by a basic group. There is not such an appealingly simple scheme in the MO model.

Discuss the 'reality' of schemes like this and the relationship between the mathematical terms occurring in a model of electronic structure. Get the views of as many experienced chemists as you can and compare them — what do *you* think?

Appendix E

Valence as Electron Spin Pairing

We have made little or no use of the spin of electrons in our descriptions of the chemical bond and yet one may find, in many accounts of valence theory, the assertion that chemical bonds are the result of the 'pairing' or 'coupling' of electron spins.

It is very obvious that most molecules are 'closed shell' — they have an equal number of electrons of both possible spins — and the two electrons responsible for a σ chemical bond certainly have opposite spin in order that the Pauli principle be satisfied. The real question is:

> Is this spin pairing a contribution to the *reason* why a chemical bond is formed or is it simply something which *accompanies* bond formation?

If it is one of the reasons for bond formation then we have made a serious error in not mentioning it, relying, as we have, on the three factors: electrostatic interactions, Schrödinger's mechanics and the Pauli principle to give a very good explanation of the chemical bond. The case for spin pairing being considered as a factor, even *the* factor in bond formation, is contained in a famous formula published by Paul Dirac in 1929. Both the nature of Dirac's formula and the date of its publication are worth commenting on.

Dirac showed that, in the hydrogen molecule, one could describe the bonding by a very simple formula:

$$E_{HH} = J_{HH} - \frac{1}{2} K_{HH} (1 + 4 \hat{S}_1 \cdot \hat{S}_2),$$

where E_{HH} is the total energy of the molecule, J_{HH} and K_{HH} are parameters with the dimensions of energy and, most importantly, $\hat{S}_1$, $\hat{S}_2$ are the 'spin operators' for the electrons forming the chemical bond in H_2. The subscripts HH have been added to E, J and K since different bonds have different values of these parameters.

In other words, this formula claims to describe the formation of the bond entirely in terms of the properties of the spin of the two electrons which form the bond — something which we have not even considered! The two parameters J and K can be chosen from experimental measurements to make the formula fit the observed energy of the bond, and this provides a model for the two-electron σ bond. It is simply a matter of finding suitable values of J_{AB} and K_{AB} for each type of bond A–B (C–H, C–C, C–O, etc.) and this formula can be extended to give a description of many σ-bonded molecules independently of the ideas that we have used.

If this is the case then we must try to check *why* the pairing of electron spins can cause such a profound change in the electron distribution of atoms and molecules.

Electrons are said to have spin not because they can be observed to be spinning since, as we have seen, only the probability distribution of their *position* in space can be calculated or measured, so it is difficult to see how one could determine whether or not an electron is spinning when we cannot even *locate* it precisely. What we can observe is that electrons behave like magnets — in addition to their charge they have a small magnetic moment which can be detected by their behaviour in a magnetic field. Magnetism is caused by electricity in motion and so the way in which an ordinary, charged body can have a magnetic field is if it is spinning. We therefore, purely by analogy, say that the (microscopic) electron has a spin.

The magnitude of this magnetic moment can be measured and so we can use a formula of magnetic theory to give some idea of the energy of inter-action between two electrons due to their magnetic moments. In 'chemical units' this energy is about 10^{-6} (0.000001) Kcals per mole for two elec-trons separated by a typical bond distance. Unlike the Coulomb interac-tion, which is always repulsive, the magnetic interaction can be attractive or repulsive depending on the relative orientation of the two magnetic moments. For comparison, the Coulomb repulsion between two electrons separated by the same distance is about 50 Kcals per mole and a typical σ-bond energy is about 100 Kcals per mole. Thus, the difference in energy between spins being paired or parallel is entirely negligible compared to the actual bond energy; the interactions between electron spins is *tiny* on the scale of bond energies.

So, either the real source of the bond energy is hidden by Dirac's equa-tion or there is some kind of occult interaction between electron spins that no-one knows about. The answer, of course, is in the way in which the *real* factors governing bond formation are hidden in the equation.

If we carry out an ordinary MO calculation of the energy of a hydrogen molecule, using the $1s$ AO on each nucleus and the two possibilities for the spin directions of the two electrons (paired and unpaired) we find, purposely using the same notation (J_{HH} and K_{HH}):

$$E_{HH} = \frac{(J_{HH} \pm K_{HH})}{(1 + S_{HH})},$$

where, this time, the formulae for the parameters J_{HH}, K_{HH} and an additional one, S_{HH}, are explicitly known. In words:

> J_{HH} is the energy of attraction between each electron to the two nuclei plus the repulsion between the two electrons.
>
> K_{HH} is the repulsion between the two electrons, each in the overlap distribution of the two $1s$ AOs.
>
> S_{HH} is the overlap integral between the two $1s$ AOs.

The overlap integral may be assumed to be less than one, so the equation becomes, approximately:

$$E_{HH} \approx J_{HH} \pm K_{HH},$$

where the only difference between this expression and Dirac's formula is that the spin-dependent term:

$$\frac{1}{2}(1 + 4\hat{S}_1 \cdot \hat{S}_2)$$

is replaced by the very much simpler ± 1. It is easy to show that, for the particular case of just two electrons, this term reduces to '-1' for the paired case (the bound ground state of H_2) and '$+1$' for the unpaired case (the repulsive state: two separate H atoms).

The upshot is that this whole formalism simply hides the real causes of chemical bonding: the nuclear attraction and electron repulsion terms in J_{HH} are due to the combined effect of Coulomb's law and Schrödinger's mechanics while the K_{HH} term includes the Pauli principle in addition.

The natural question to ask now is:

> Why, when it is so simple to see that the idea of spin pairing as a *cause* of chemical bonding is such a red herring, was this idea so pervasive?

There are probably many factors involved here but the clue to one of them is in the *date* of the publication of this formula. In 1929 there was no real possibility of computing the parameters J_{HH} and K_{HH} for even the $1s$ AOs and certainly not for the $2s$ and $2p$ AOs used for bonds involving other atoms; it is quite routine to use computers to compute these quantities today, but in 1929?

It should be said that the use of semi-empirical formulae to attempt to systematise and classify phenomena is a well known and perfectly acceptable method in physical science. Indeed when the laws underlying the phenomena are unknown this may be the only way to proceed, as we see in sub-nuclear particle physics. But, when the laws are known and the technical means of solving the resulting equations are available, these methods become redundant. There is a more thorough look at electron spin in Appendix H.

Chapter 6

Lone Pairs of Electrons

*The theory developed so far explains only the bonding
pattern and shapes of molecules whose electronic structure
consists of just two types of substructure: atomic cores and
single bonds. This is not enough. The idea of a 'lone pair'
of electrons is all-pervasive as a factor in the theory of the
shapes and reactivities of molecules. It proves to be just as
necessary to understand the effect that bond formation has
on the 'non-bonding' electrons of atoms as it is to have a
theory of how chemical bonds form.*

Contents

6.1 Why are Not All Electrons Involved in Bonding?

So far we have considered only those molecules in which the electrons involved in forming a bond are:

- All the electrons of each atom (H_2 and, arguably, 'He_2') or;
- Well separated in energy from any other electrons of the bonded atoms (LiH and CH_4, where the atomic cores of $1s$ electrons in the lithium and carbon atoms are undisturbed by molecule formation).

It is straightforward to show[1] that the theory also works for the stable hydrides of the atoms between lithium and carbon in the periodic table: BeH_2 and BH_3.

However, the bonding in the remaining stable hydrides of the elements in this row of the periodic table resists our simple theory:

- Nitrogen has an atomic core of two $1s$ electrons as usual and five electrons in its valence shell, suggesting a hydride of formula NH_5.
- Similarly, oxygen and fluorine with six and seven valence electrons respectively, might be expected on the basis of our theory to form H_6O and H_7F.

In fact, of course, the stable hydrides of these atoms are NH_3, H_2O and HF, each having an even number of H atoms less than the theory so far would predict. That is, in each case there is an *even* number of electrons from the central atom which are not involved in forming chemical bonds. This is a new phenomenon which we must try to explain and take on board.

First of all, think of the 'physics' of what is happening as a central atom is surrounded by more and more (hydrogen) atoms:

- Remember that the bonding pairs of electrons repel one another, the approaching atoms' atomic cores also repel each other. The more 'bonds' the more repulsion there will be between those bonds to offset the attraction by possible bond formation.
- Also recall the effect of the Pauli principle: the more electrons there are 'squashed' into a small region of space, the more the Pauli principle will resist it because there simply is not room for the constraint of just two electrons per orbital to be satisfied.

---

[1] Try this yourself.

- Finally, recall the energy levels available to electrons in the *separate* atoms. These levels form 'shells' with energies proportional to $(-1/n^2)$, as we outlined in Section 2.1 on page 35. The gaps between these levels present difficulties if one tries, for example, to form atomic negative ions.[2]

All of these three effects are pointing in the same direction: there may well be a limit to the number of electrons of a given atom which can be shared to form chemical bonds.

The reason why the number of unshared electrons is, generally, an *even* number is that it is relatively unusual for chemically stable molecules to have an odd number of electrons in total. A molecule (or an atom) with an odd number of electrons must have, in our orbital model, an orbital containing just a single electron. This necessarily means one of two things:

(1) The 'odd electron' is in an orbital higher in energy than the other, paired electrons.[3]
(2) The odd electron is in an orbital with similar energy to the other, paired electrons; i.e. there is a 'hole' in that orbital which can be occupied by another electron.

Whichever of these conditions is true, it means that the odd electron will be chemically reactive, in the sense that it will either be easily lost from the high-energy orbital or be easily paired by another electron filling the hole in the lower-energy orbital.

So, for *chemically* stable molecules we expect two things in general:

- Atoms from the 'right-hand side' of the periodic table will tend to form fewer chemical bonds than the number of electrons in their valence shell.
- The number of chemical bonds formed by an atom with a number (N, say) of electrons in the valence shell will tend to be given by:

$$N - 2n,$$

where the number n is a small integer: 0, 1, 2, typically. Thus, for carbon $N = 4$, $n = 0$ the hydride is CH_4; for nitrogen $N = 5$, $n = 1$,

---

[2]That is, isolated negative ions, not ions in solution, which are a different phenomenon altogether.
[3]Why is this so?

NH_3; for oxygen $N = 6$, $n = 2$, H_2O, and for fluorine $N = 7$, $n = 3$, HF.

Although these electrons are not involved in bonding in molecules, they have profound effects on the shapes and reactivities of the molecules in which they are present.

6.2   What is a Lone Pair?

We have established good reasons for thinking that there will be valence-shell electrons not involved in bonding. It is common for electronic structures of this kind to consist of an even number of electrons and, what is not at all obvious, that these electrons behave towards other molecules as if they are grouped together in pairs. This is what we shall call 'lone pairs' of electrons, and our purpose in this section is:

- To justify the name 'lone pair' by looking at the distributions taken up by non-bonding electrons.
- To account for the characteristic reactions of such lone pairs, or, rather, the reactivities of molecules containing lone pairs.

The existence and lack of chemical reactivity of the inner-shell electrons in heavy atoms do not need a specific justification, independent of the fact that they are so tightly bound to their 'parent' atoms. So, from this point of view, the electrons occupying the $1s$ AOs of the N and O atoms are not lone pairs. This choice fits into our earlier definition of 'atomic cores'. The electrons in atomic cores do not count as lone pairs.

What can we say about the way in which these non-bonding valence electrons will be distributed around their parent atom? The factors involved are, as usual:

(1) The electrostatic interactions amongst charged particles (electrons and nuclei or combinations like atomic cores).
(2) The Pauli principle and its effect on orbital occupation.

If we consider an electronic structure composed of atomic cores, a bond-pair of electrons and some non-bonding valence electrons, we can see what might happen:

(1) In the separate atom the electrons are distributed in an unpolarised way.

(2) The formation of chemical bonds establishes polarised regions of electron distribution.

(3) These polarised regions themselves polarise the distribution of the non-bonding electrons. The overall effect will obviously be to push the non-bonding electrons to the side of the atom furthest away from the bond pairs: to the 'back' of the atom.

(4) These non-bonding electrons will have repulsions among themselves, of course, and they will try to get as far away from each other as possible, consistent with avoiding the electrons in the bond.

(5) The Pauli principle constrains these — like all — electrons to occupy orbitals in pairs.

The best way to illustrate this process is by way of an examination of a particular example. Since we have spent so much time discussing methane let's look at its next-door neighbour, ammonia, since the problem is simplified by the fact that there are only two valence-shell electrons not used in bonding.

6.2.1 *The ammonia molecule*

If we go through a similar procedure for ammonia as we did for methane, using only the approach of three hydrogen atoms, we obtain:

(1) Three bonds of the same general type as the ones in LiH and CH_4. Obviously they differ in detail but they are qualitatively similar.

(2) A pair of electrons — non-bonding electrons, say — remaining; based on their parent nitrogen atom and of, as yet, unknown distribution.

(3) An unknown shape for the molecule.

First things first: what do we expect for the distribution of the non-bonding pair of electrons? Unlike the bonding electrons, their distribution will be only weakly polarised by the approach of the hydrogen atoms. The dominant effect will be the repulsions between them and the other, 'nearby' electrons in the molecule: the bond pairs. These mutual repulsions will determine both the distribution of the lone pair of electrons and the overall shape of the molecule. These four electron pairs will orientate themselves so as to minimise their mutual repulsions. As usual electrons will try to get as far away from each other as is possible, consistent with the bonding scheme.

Thinking about this for a while[4] shows that there are two possibilities for the linked problems of the distribution of the lone pair and the overall shape of the molecule:

- The three N–H bonds will lie in a plane with the bonds mutually at 120 degrees; i.e. the hydrogen atoms will form an equilateral triangle with the nitrogen atom in the middle of the triangle. The lone pair of electrons will be distributed symmetrically above and below the plane of the molecule. This way the bonds are as remote as possible from each other and 'fairly' remote from the non-bonding electrons.

- The three N–H bonds will form a symmetrical pyramid with the hydrogen atoms 'below' the nitrogen atom, and the non-bonding electrons will take up a distribution predominantly on the side of the nitrogen atom, away from the pyramid of hydrogen atoms: 'above' the nitrogen atom. This way the bonds are not as remote from each other as in the previous option, but they are further away from the lone pair.

But which is it? The purely pictorial, qualitative methods used so far do not enable us to give an answer to this, basically quantitative, problem.[5] There are two routes open to us:

(1) The structure of the ammonia molecule is *known*; why not just take this and try to explain why it is like it is?

---

[4]Try thinking about this for yourself.

[5]Notice that the possibility of the molecule not being *symmetrical* — N–H bonds of different lengths and/or H–N–H angles different — is not being considered. Is this justified? Are we storing up trouble here for the future when we have to think about molecules with more complex structures?

(2) Using the quantitative methods of quantum theory we can *calculate* the energy of each of these possible structures and see which is the lower.

In either event, we need to extract some general qualitative principles from the results of this and similar molecules.

The advantage of method (2) is that — unlike method (1) — it gives us *both* the required shape *and* the actual distribution of the pairs of electrons. So we will use method (2) and, of course, check that it does give the same answer for the shape of the molecule as method (1) ('belt and braces').

Here, then, are the calculated distributions — orbitals, actually — and energies for the two types of electron pair in the NH_3 molecule. First, the planar case:

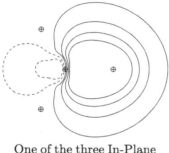
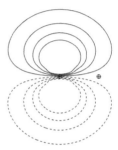

One of the three In-Plane   The Out-of-Plane
Bond-Pair Orbitals   Lone-Pair Orbital

(Total energy of this shape −55.86942 atomic units)

and now the pyramidal case:

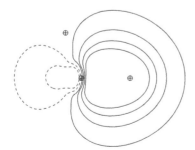
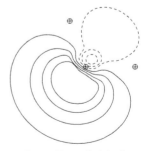

One of the three Bond-Pair Orbitals   Lone-Pair Orbital

(Total energy of this shape −55.87220 atomic units)

Since the pyramidal case is not at all easy to appreciate from these diagrams, the two orbitals are repeated below in 3D perspective; a single contour in each case.[6]

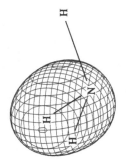 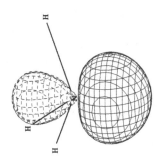

There are several points of note here:

- The pyramidal shape is the lowest in energy and the experimental measurements show that the molecule is indeed pyramidal. But the energy difference is small. Atomic units are rather large; if we convert to conventional 'chemical' units the difference between the two shapes is only about $7.3\,\text{kJ}\,\text{mol}^{-1}$.[7]
- The shape of the bond-pair orbitals is very similar in both cases.
- The shape of the lone-pair orbital is quite different in the two cases:

  - In the planar case, it looks exactly like a $2p$ AO.
  - In the pyramidal case, it is rather similar to the HAOs used to form the bonding MOs.

- A point to note in particular is that the lone-pair orbital is 'fatter' than the bond-pair orbitals; presumably because there is no attractive H atom to pull the distribution out. Also notice, in the 3D diagram, the larger 'tail' on the lone-pair contour. Overall, the lone pair of electrons tends to occupy a larger region of space than the bond pair, since it is only attracted to one atomic core.

Before going on to draw some general conclusions about the rôle of lone pairs in general molecules, we can look at the next obvious example: the water molecule, H_2O.

---

[6]The point of view is changed slightly to get a simpler appearance.

[7]So, presumably, it is quite easy to turn the molecule 'inside out' like an umbrella.

6.2.2 *The water molecule*

Without going through the same detailed treatment of the water molecule, we can look at the four non-bonding electrons in H_2O. With the experience of NH_3 in mind we can visualise three possibilities for the electronic structure of H_2O:[8]

(1) A linear molecule with parallel O–H bonds and two mutually perpendicular '$2p$-like' lone pairs.
(2) A bent molecule (necessarily planar) with two *different* lone pairs: one $2p$-like and one in the same plane as the two H atoms, more like the lone pair in NH_3.
(3) A bent molecule with the same qualitative 'shape' as CH_4 and NH_3: an approximate tetrahedron with only two H atoms at the apices and two lone pairs 'pointing' at the other two apices.

There is some dispute about which of the last two is the better description of the electronic structure of the molecule, since both would predict a bent molecule which is found experimentally. The dispute arises because of the different predictions which the two structures would generate when involved in so-called 'dative' bonds. For the moment, for the sake of simplicity, we assume that the 'tetrahedral' model works well for the trio CH_4, NH_3 and H_2O. Contour diagrams of the bond- and lone-pair orbitals in H_2O are very similar to those given above for NH_3, and so are not given here. Notice that the electronic structure of the (linear) molecule hydrogen fluoride (HF) has not been discussed here; this is deferred until we have looked at diatomic molecules a little more carefully.

So, in answer to the question which was posed by the title of this section, 'What is a Lone Pair' we can give the following (provisional) answer:

> A lone pair of electrons in a molecule is a pair of electrons which occupy an atomic orbital (AO or HAO) in the valence shell of an atom, and which are not involved in the bonding of the molecule. A lone pair differs from a pair of electrons in the atomic core of an atom by being sufficiently loosely bound to be capable of being involved in the chemistry of the molecule.

---

[8]Remember that *any* triatomic molecule is planar.

6.3 The Shapes of Simple Molecules

Having discussed lone pairs of electrons, we are now in a position to try to draw together the ideas developed so far to see how they may be applied to answer the question, 'why do molecules have the shapes that they do?' First of all, recall the reasoning we used to explain the shape of the methane molecule in Chapter 5:

(1) Attraction between one H nucleus and the two electrons of another bond is 'two terms', giving a total of four terms for this effect.
(2) Repulsion between two bond pairs of electrons is four terms: each of two electrons repelling each of the other two.
(3) Repulsion between two H nuclei of separate C–H bonds is one term.

The conclusion here was that, since the repulsive terms outnumbered the attractive terms by five to four, there was a net repulsion between bonds. This 'conclusion' was reached without considering the actual *sizes* of the terms and may give the right answer for methane by sheer coincidence.[9]

When molecules contain lone pairs of electrons as well as bond pairs, the above considerations will still apply, but, in addition, we will have to consider the interactions between bonds and lone pairs and between one lone pair and another if the molecule has more than one lone pair. Again, we can look at this problem purely qualitatively, applying the ideas of Chapter 5.

For the interaction between a bond and a lone pair:

(1) Item 1 in the list above is modified because there is no nucleus 'at the end of' the lone pair of electrons. So, there are only two terms this time.
(2) Item 2 stands as it is: two pairs of electrons, four terms.
(3) Item 3 does not apply: no nucleus in the lone pair to repel the nucleus in the bond pair.

Net effect: four repulsive terms, only two attractive terms. So a repulsion between a lone pair and a bond pair might well be stronger than between two bond pairs.

---

[9]It is hard to see anything leading to a shape for methane which is not tetrahedral!

Now think about the interactions between two lone pairs:

(1) Item 1 in the list no longer applies: there are no nuclei to attract the electrons of the lone pairs.
(2) Item 2 again stands: two pairs of electrons, four terms.
(3) Item 3 does not apply again.

Net effect: four repulsive terms and *no* attractive term. Repulsions between lone pairs will be even stronger than the repulsions between a bond pair and a lone pair.

Remember, this is all purely qualitative; we have no quantitative data to use with which to clinch these arguments. But, based on this evidence we can say, provisionally, for the moment:

> A major factor influencing the shape of molecules is the inter-action between bonds and lone pairs which share a common atom. Qualitatively it looks as though the net interactions amongst these structures are *repulsive* and the magnitudes of the repulsions are in the order:
>
> LP with LP > LP with BP > BP with BP,
>
> where 'LP' means a lone pair and 'BP' means a bond pair of electrons.

This provisional conclusion will be reinforced by some numerical calculations later; for the time being it will be regarded as justified by the arguments just presented. These rules are well known as the valence-shell electron-pair repulsion (VSEPR) model, although it is clear that the repulsions among the bond and lone pairs are not the only forces to be considered in looking for an explanation of the shapes of simple molecules.

6.3.1 *The water molecule — again*

With these provisional conclusions, we can now look at the water molecule again and see what can be said about its shape. As we noted above there are three possibilities: linear with two '2p-like' lone pairs; bent with one '2p-like' lone pair and one polarised (HAO) lone pair; bent with two polarised (HAO) lone pairs. Two of these possibilities involve 2p-like lone pairs, which we can see are an exception to our rule generated above.

The (triatomic) water molecule is necessarily planar,[10] and in the two cases in question we have a lone pair whose distribution is symmetrically disposed above and below the plane of the molecule. It is easy to see that, in such an arrangement, the bonds can make any angle with each other without affecting their repulsions with such a '$2p$-like' lone pair.[11] It is therefore the remaining (σ) bonds and lone pair(s) which will determine the shape of the molecule. Each of the three cases has two O–H bond pairs and two lone pairs of course; they only differ in the *type* of lone pair structure:

- A linear molecule with two $2p$-like mutually perpendicular lone pairs.

  The lone pairs have no effect on the shape, so the remaining bond pairs simply minimise their mutual repulsions by getting as far from each other as possible, making the molecule linear.
- A bent molecule with one $2p$-like lone pair and one (in-plane) polarised HAO lone pair.

  The $2p$ lone pair does not affect the shape, so, as above, the shape is determined by the repulsions amongst the remaining two bond pairs and one lone pair. These three take up a shape in the plane of the molecule and the angles between the three of them are determined by the relative strengths of the repulsions. If the repulsions were all the same the shape would be trigonal: angles of 120 degrees between the three electron pairs. Since the BP–LP repulsions are larger than the BP–BP repulsions, the angle between the two O–H bonds will be smaller than the angle between the lone pair and each bond. So, the bent molecule will have an angle between the O–H bonds of less than 120 degrees.
- A bent molecule with two polarised HAO lone pairs.

  If the BP–BP, LP–BP and LP–LP repulsions were all the same, the molecule would have the four electron pairs tetrahedrally disposed in exactly the same way as the methane molecule. That is, the H_2O molecule would be bent with an angle of about 109.5 degrees. Since we know the relative strengths of the different repulsions, it is easy to see that, as in the previous case, the angle between the O–H bonds would be

---

[10]This fact is related to the reason why bar tables often have three legs: they can be placed on any surface without wobbling.

[11]Be sure that you are convinced that this is true. Make a model and try it out; open and close the bond angle and see that the repulsions between a π lone pair and bonds in the plane are not affected.

somewhat less than this, and the angle between the lone pairs somewhat more.

The water molecule is found experimentally to be bent with an angle of about 104.5 degrees between the two O–H bonds, suggesting that the last of the three possibilities above is the closest to the truth.

In these kinds of contexts it is important to recognise that the ideas presented here only apply to an *isolated* water molecule. Thus, in some chemical reactions the water molecule forms complexes with metal ions in which the two O–H bonds are symmetrically disposed to the metal atom, suggesting the second of the above models. But saying that the third is more likely does not exclude the possibility that the approach of another atom or molecule cannot *change* the electron distribution. Indeed, the very idea of chemical reaction presumes that electron distributions are changed on close approach, otherwise new bonds would never form.

6.4 'Reactions' of Lone Pairs

If we think about the electron distribution in the NH_3 or H_2O molecule and, in particular, the distribution of the lone pairs — the valence shell electrons not involved in bonding — it is clear that these electrons will be an irresistible attraction to certain types of atom, ion or molecule. Pairs of electrons, relatively localised in space and only attracted to a single nucleus, just cry out to attract any positively-charged species in their vicinity. The negatively-charged lone-pair distributions illustrated above for NH_3 and H_2O will certainly attract, for example, a hydrogen cation H^+ or, in general, *any* positively-charged ion like Na^+ or Cu^{2+}. If we concentrate on the hydrogen cation, where the situation is not complicated by other electrons, it is quite clear that, for example, the lone pair of the NH_3 molecule will probably 'capture' the proton and form a new molecule NH_3–H^+. Just take a look at the lone pair 3D contour on page 113; it looks like a negatively-charged 'web' waiting to catch a positively-charged 'fly'. What is more, the repulsions between the existing bond pairs and this new structure, formed from the lone pair and the captured proton, will cause the shape of the molecule to change: the angles between the N–H bonds will change. The most straightforward view might therefore be that this new entity would have the three original (equivalent) N–H bonds *plus* a new way of bonding a proton to a nitrogen atom. However, as we shall see later in more detail, this 'new' entity is nothing more than the familiar (tetrahedral) NH_4^+ ion in which all *four* N–H bonds are the same. Clearly there is some explaining to do about this type of structure which, for the moment, we shall have to defer.

In fact, this type of structure is very common in chemistry and involves a greater range of bonds than the N–H type. The importance of 'dative bonds' — as these structures are called — deserves much more detailed consideration. This investigation is deferred until Chapter 10, after we have looked at another new type of valence structure: the double bond.

6.5   A Working Summary

This might be a good place to summarise what we have found so far, even though we have only understood just two types of valence-electron structure: bond pairs and lone pairs. The general principles are:

- A chemical bond is formed when the energy of a pair of electrons (one from each partner in the bond) can be lowered by being shared between the attractive force field of two atomic cores simultaneously, rather than each electron being attracted solely by its 'own' atomic core.
- When an atom is involved in several bonds with different atoms there is always repulsion between these bonds (between the electrons of the pairs and between the atomic cores of the bonded atoms).
- How many atoms a given atom can bond to depends on the balance between these two things:

  (1) The energy lowering due to the sharing of electrons.
  (2) The energy raising due to the repulsions among the bonds formed.

What is found is that, at least for the atoms H through Ne, atoms with a small number of valence electrons — H through C — can share *all* their valence electrons to form bonds, while the remaining atoms — N through Ne — do not share all their valence electrons but form lone pairs with some of them. The reasoning for this particular state of affairs is summarised on page 108, and we shall see later in Chapter 13 that the same reasoning under different circumstances leads to quite different conclusions.

6.6   Assignment for Chapter 6

The net repulsions amongst the various electron pairs in a molecule can be more easily understood by simply thinking about the ways in which a series of charges would arrange themselves if they were confined to the surface of a sphere. The shapes which these arrangements form, are useful starting points for the use of the VSEPR model.

(1) Think about the various possibilities; what are the arrangements for two, three, four, five and six equal charges confined to the surface of a sphere but free to move over that surface?
(2) Now introduce the idea that the charges are not necessarily equal and consider labelling them either BP (for a bond pair of electrons) or LP (for a lone pair of electrons). Try a few examples, e.g. for four charges try two LPs and two BPs, and for five charges try three BPs and two LPs. Use the 'rule' from page 117 to guide your thinking.
(3) Now predict the shape of the $C\ell F_3$ molecule.

One way of interpreting the dipole moments of molecules is to use the data from a series of molecules containing bonds between the same pairs of atoms, and see if it is possible to define the dipole moment of each *bond* and assume that this quantity is environment insensitive. This enables the prediction of the moments of other molecules containing these bonds. Discuss why the existence of lone pairs might upset this scheme — look at the diagrams of the HAOs of H_2O in Section 6.22 on page 115.

Chapter 7

Organic Molecules
with Multiple Bonds

*So far, we have looked at only three types of electronic
substructure occurring in molecules: atomic cores, single
bonds and lone pairs. There are a lot more to be investigated. We must, therefore, continue our investigation of
the different types of environment-insensitive electronic
substructures in molecules. In some ways, a consideration
of the presence of multiple bonds in polyatomic molecules
is easier and more direct than the approach of the problem
through diatomics. The rather exceptional high (cylindrical) symmetry of diatomic molecules complicates matters
rather than simplifying them. There is, for example, no
diatomic molecule in which there is a typical double bond
of the type found in thousands of organic molecules: two
bond pairs. Also, of course, along with lone pairs, it is
the existence of multiple bonds in organic molecules which
gives them their interesting and valuable chemical properties. We must, therefore, take a careful look at double and
triple bonds in organic molecules.*

Contents

7.1   Double and Triple Bonds

Although it is reasonable to start an account of the *theory* of molecular electronic structure with an account of the single (electron-pair) bond, molecules containing only single bonds — saturated molecules, as they are called in organic chemistry — do not have a very interesting chemistry. Think of the methane molecule or any other saturated hydrocarbon: all they do is burn, which is useful but not very interesting. We have seen that lone pairs do give a molecule some interesting chemical properties. The other main electronic substructure which gives a molecule some chemically-interesting features is the existence of multiple bonds: two atomic cores sharing more than two electrons. We need to look at this phenomenon if we are to get to grips with chemistry.

There are new structures to be found even in some of the most familiar organic molecules. A close look at the simplest will enable us to examine the phenomenon of multiple bonds: two atoms sharing four or six electrons without violating the Pauli principle. It is common in descriptions of the quantum theory of valence to begin with a fairly thorough discussion of the bonding in diatomic molecules, for the apparently logical reason that they are the simplest molecules. Although it is true that they are simple, in the sense of containing only two atoms, and therefore can only contain bonds between those two atoms, in fact, the multiple (double and triple) bonds in those diatomic molecules which contain them are far from typical because:

- There is no diatomic molecule which contains a 'straightforward'[1] double bond. That is, for example, among the doubly-bonded diatomics formed

---

[1] We shall see what this rather vague term means as we go along.

from the atoms Li to F, in none of them is the double bond formed from *two electron pairs* of the type we have been discussing so far. The most familiar one, O_2, for example, has a complicated open-shell structure (it is magnetic) and the highly reactive, unstable C_2 molecule is just as untypical.

• The *energetics* of the molecules containing triple bonds are not at all typical of multiple bonds in the familiar organic polyatomic molecules. The N_2 molecule, for example, undoubtedly contains a triple bond between the two atoms, but, if we compare the *strength* of that triple $N\equiv N$ bond with the strength of a single $N-N$ bond in a polyatomic molecule and then compare these differences with, for example, the $C\equiv C$, $C=C$ and $C-C$ bonds in typical organics, we find clearly different trends.

Bond	$N-N$	$N\equiv N$	$C-C$	$C=C$	$C\equiv C$
In What?	NH_2-NH_2	N_2	alkanes	alkenes	alkynes
Energy (kcal mol^{-1})	39	226	83	146	199
Difference		187		63	53

In the familiar organic molecules the energy of the *double* $C=C$ bond (146) is less than twice that of the single bond ($2 \times 83 = 166$), and the energy of the *triple* $C\equiv C$ bond (199) is much less than three times the energy of the single bond ($3 \times 83 = 249$), and less than one and a half times the energy of the double bond ($1.5 \times 146 = 219$). In other words, each successive bond added to the original $C-C$ single bond contributes *less* to the total binding energy.

In contrast, the triple bond in the N_2 molecule is nearly *six* times as strong ($226 \div 39 = 5.8$) as the single $N-N$ bond in hydrazine. A similar condition holds for the untypical double bond in the oxygen molecule: the $O-O$ single bond in H_2O_2 (hydrogen peroxide) has an energy of 34, while the double bond in O_2 has a total energy of 119, almost three times the strength of the single bond.

The sequence found in the organic molecules is much the most common: the single bond is the stronger of the two in a double bond and the strongest of the three in a triple bond. Note that the energy addition in going from two bonds to three ($199 - 146 = 53$) is closer to that in going from one bond to two ($146 - 83 = 63$) than the bond energy of the single bond (83). We shall be able to explain this fact during this chapter.

There is clearly something special about the bonding in *diatomic* molecules containing multiple bonds.

Since our *main* aim in investigating the electronic structure of molecules is to develop a theory of the environment-insensitive substructures within molecules, it is sensible to look at the 'typical' case of multiple bonding in polyatomic molecules, and to defer the theory of the bonding in diatomics — and an attempt at an explanation of their anomalies — until a later chapter.

7.2   The Possibilities

Organic molecules provide a much more varied and rich field of possibilities for multiple bonding than simple diatomics. We shall see that there are double and triple bonds in organic molecules and the additional possibility of many such bond schemes in a single molecule, because of the size and complexity of organic molecules containing the bonding between carbon atoms.

We can start by using a very simple model and ask an obvious question: why is a double bond between two carbon atoms not simply the result of the formation of *two* bonding pairs of electrons, each of which is a combination of two hybrid atomic orbitals, one from each carbon atom?

> That is, is the ethene molecule just an ethane molecule from which two hydrogen atoms have been removed, and the 'unused' HAOs combined to form a second bond in much the same way as the existing C–C bond in ethane?

Long before the existence of any quantum theory of the chemical bond, this was precisely the assumption which was made, and very reasonable it seems too. If one pictures the carbon atom at the centre of a tetrahedron, then:

- Methane (CH_4) has the four hydrogen atoms at the vertices (corners) of a tetrahedron (correct).[2]
- Ethane (CH_3–CH_3) has one hydrogen from each atom removed and the two tetrahedrons[3] are joined at a shared *vertex* (very nearly correct[4]).

---

[2]Ignoring the obvious fact that the molecule is always vibrating!

[3]I use tetrahedrons instead of tetrahedra throughout.

[4]Not all the bond angles are exactly the same.

- Ethene ($CH_2=CH_2$) has two hydrogens removed from each atom and the two tetrahedrons are joined at a shared *edge* of two tetrahedra (to be tested).
- Finally, ethyne ($CH\equiv CH$) has three hydrogens removed from each atom, and the two tetrahedrons are joined at a shared *face* (to be tested).

Certainly this scheme[5] has a very attractive feel about it, and it does *rationalise* some of what we know empirically about the possible bond types in organic molecules. It even provides a neat explanation of why it is hard to 'rotate' a multiple bond! The question for us, however, is 'does it square with what we *now* know about the properties of multiple bonds in organic molecules?'

Other than its empirical formula and its connectivity, the most basic things about an organic molecule are:

- Its 3D shape.
- Its reactivity.

What does the 'tetrahedral model' (as we might call the above scheme) tell us about these properties?

- Ethene is found experimentally to be a planar molecule and ethyne is linear. Both of these facts are predicted perfectly by the tetrahedral model, which is easy to see by manipulating a pair of tetrahedrons. But the H–C–H bond angle in ethene is very close to 120 degrees, and not the tetrahedral angle of about 109.5 degrees. This is suggestive but hardly decisive.
- The chemistry of both ethene and ethyne are both *very* different from that of ethane. Both of the molecules containing multiple bonds show many reactions in which the two carbon atoms remain attached by a single bond, and the other bond or bonds between the two C atoms are transformed into some quite different structures. This is obvious from the differences in length of the descriptions of the chemistry of the alkanes and of, for example, the alkenes in any organic chemistry text. This immediately suggests that the two bonds joining the atoms in an alkene (or the three bonds in an alkyne) are not at all equivalent as the tetrahedral model would suggest, but that there is *one* alkane-like bond and the other bond or bonds are different from this one.

---

[5]This was proposed by van 't Hoff, one of the joint discoverers of the 'tetrahedral carbon atom'.

Later on, when we come to look at the properties of aromatic and conjugated molecules — molecules with *alternating* single and double bonds — we shall see that the attractive simple tetrahedral model is completely unable to explain the range of new phenomena which these molecules exhibit.

With this evidence in mind and the experimental evidence of the bonding in diatomics containing multiple bonds, we can take a look at some simple organic molecules containing double bonds.

7.3   Ethene and Methanal

The simplest and most familiar organic molecules containing multiple (double, actually) bonds are ethene (C_2H_4)[6] and methanal (CH_2O).[7] Both of these molecules contain a double bond between the two non-hydrogen atoms. We need to use the methods developed so far to provide a picture of the bonding in these molecules.

First of all, consider each of the carbon atoms in ethene and the carbon atom in methanal. The atoms to which these carbon atoms are bonded (i.e. adjacent to in the molecule) are found experimentally to be all *in the same plane* and disposed about the carbon atom approximately symmetrically, with H–C–H and H–C–O bonds of around 120 degrees. Remembering what we said about the relative orientation of bond pairs of electrons — they adopt an orientation which minimises their mutual repulsions — this fact suggests that there are three *similar* bonds from each of these carbon atoms to their nearest neighbours because the minimum energy of three similar electronic structures in mutual repulsion is just *pointing at the vertices of an equilateral triangle*, i.e. forming angles of 120 degrees.

The question which remains, however, is:

> How do we explain that the two C–H bonds attached to these carbon atoms are obviously *not* the same as the C=C bond in ethene and the C=O bond in methanal?

It looks as if the shapes of ethene and methanal are determined by just the single bond structure of the molecule, in a very similar way to the way in which the shape of ethane is determined by the repulsions between

---

[6]Formerly ethylene; its polymer is still known as polyethylene or simply polythene.

[7]Formerly formaldehyde and still known commercially by this name when used (e.g. by the artist Damien Hirst) as a preservative.

the single C–H and C–C bonds *without even having to consider the fact that ethene and methanal contain a double bond.* That is, the two electrons which form the second bond of C=C and C=O seem to have no effect on the shape of the molecule *even though, like all electrons, they must be repelling any other electrons in the molecule!*

All electrons repel one another, so there must be an explanation of the bonding and shapes of these two molecules which takes account of the facts that:

• All the electrons are attracted to each atomic core.
• All the electrons repel each other.

There must be some way in which the electrons of the second bond of the double bond repel the electrons of the C–H bonds and the electrons of the first bond of the double bond *which does not affect the shape of the molecule determined by the mutual repulsions of the 'single bonds'.* This is just a rather awkward way of saying that the 'second bond' electrons must repel the single bond electrons (approximately) *equally.*

If we recall the reasoning behind the shape of the ammonia molecule in Section 6.2.1, the answer can be seen quite quickly:

> The electrons in the second bond of the double bond must be equally distributed *above and below* the plane containing the atoms of the molecule and have very little density in the plane of the molecule.

Because:

• If the electrons are distributed out of the plane of the molecule, they will be relatively remote from the other electron pairs and so not have a large effect.
• If they are *equally* (symmetrically) distributed above and below the plane of the molecule, the net effect of the repulsions will cancel out and leave the shape undisturbed.[8]

So, what we have to do now is to find a way of describing the electronic structure of the second bond in a double bond which is consistent with these simple conclusions and which will account for the more interesting chemical properties of double bonds.

---

[8] Make sure that you believe this!

7.4   The Double Bond in Ethene and Methanal

Let's assume that each of the lowest-energy electron pair in the C–H bonds and the C=C bond in ethene can be described by the method we have used for ethane: a localised MO formed by combining a HAO on each of the constituent atoms of the bond.

- Each HAO on the carbon atom is a polarised AO which can be expressed as a linear combination of the unpolarised AOs ($2s$ and $2p$) of the atom.
- The HAO on the hydrogen atom is a (slightly) polarised hydrogen $1s$ AO.

Then the simplest electronic structure which is symmetrically distributed above and below the plane of the molecule is a localised MO formed from an HAO on each carbon atom which is itself symmetrical in the plane of the molecule. Such an HAO must be a combination of p-type AOs since the s-type AOs on the carbon atom, although being symmetrical, have most of their distribution *in* the molecular plane.

We therefore assume that the second bond of the C=C double bond is of a quite different structure from the first:

> The second bond of a double bond is described by a linear combination of an HAO on each of the bonded atoms which is of p-type, i.e. a MO which changes sign with respect to reflection in the plane of the molecule.

The electron distribution due to two electrons occupying this MO is obtained from the *square* of this MO and is, of course, all positive but symmetrical above and below the molecular plane, as we require. A natural question to ask about this new type of electronic structure is: 'are the HAOs which are used to form the MO *polarised* in the same way as the ones involved in the more familiar single bonds?' The answer is 'in general yes, but the *amount* of polarisation is very small.' If we look at the possibilities for polarising a $2p$-type AO while retaining its characteristic feature of changing sign with respect to reflection in the plane of the molecule, there is only one type of thing which can happen:

> The $2p$-type AO can be made to 'lean'. That is, the dumbbell-shape can be symmetrically slightly 'folded' (presumably) in the direction of the atom in which it is involved in bonding.

Careful calculations show that this does, in fact, happen in both the ethene and methanal molecules, but the amount is *very* small, too small to be detectable on the sort of simple contour diagrams we have been using. The effect does show up in the calculations; however, it appears as a small amount of a $3d$-type AO (from around 0.2% in ethene to about 0.4% in methanal) being mixed into the predominantly $2p$ AO (the rest: 99.8% and 99.6% respectively). The reasons why this effect is so small can be interpreted as the same as the reason why this structure does not affect the shape of the molecule:

- The electron in the $2p$ AO is more remote from the atomic cores of the other atoms.[9]
- There is a fairly symmetrical distribution of atomic cores around the $2p$ AO.

So, from now on, we can drop the 'H' of 'HAO' if we choose to do so and simply say that:

> The second bond of the double bond can be represented as a combination of a $2p$-type AO on each of the bonded atoms which is perpendicular to the plane of the molecule.

Having said all this, it is time to look at the results of some actual calculations to make sure that we are on the right track and not just spinning reasonable-sounding ideas out of our heads. Before presenting the actual results, it is useful to establish the nomenclature for these systems. Rather than having to say 'the lowest-energy bond of a double bond' and the 'new structure', we can use what are now the standard names for these structures. The terminology is obtained by comparing the appearances of the contour diagrams of the MOs for the two types of system with those of AOs.

7.4.1 *Sigma (σ) and Pi (π) notation in planar molecules*

In our discussions in earlier chapters we have only considered $2p$ AOs as part of the contribution of the atomic electron distributions in forming polarised HAOs prior to the formation of an electron-pair bond. These electron-pair bonds have all been (approximately) cylindrically-symmetrical around the bond axis; if we 'look' at them

---

[9]Remember, the electrostatic energy falls off rapidly with distance.

'end-on' they have an (approximately) circular appearance. But, we have now seen that there are other possibilities. There are situations in which it is perfectly possible to form an electron-pair bond in which the electron distribution does not have this 'end-on circular' property.

To avoid a lot of unnecessary descriptions, we can introduce some suitable nomenclature. If we imagine a bond between two atoms and look at it 'end-on' — along the internuclear axis — the familiar $2s$ and $2p$ AOs centred on each nucleus fall into two distinct classes:

(1) Those which 'look like' an s orbital, i.e. the contours are circular. There are just two of them: the actual $2s$ AO and the $2p$ AO lying along the bond axis (the dumbbell shape is hidden from our point of view).
(2) Those which do not: the two remaining $2p$ AOs which are in two mutually perpendicular directions both perpendicular to the bond axis. These two AOs still have the familiar '$2p$' appearance from our chosen point of view.

Thus the $2s$ and just one of the set of three $2p$ AOs are suitable for describing an HAO polarised along the bond axis (in either direction) and the other two $2p$ AOs are not. It is useful to have a notation which distinguishes between the $2p$ AOs when they are involved in describing localised electron-pair bonds by making this 'end-on' view a bit more formal:

> Orbitals which do not change sign when reflected in a plane containing the bond axis are known as σ orbitals, and orbitals which do change sign are called π orbitals.[10]

Thus, among the orbitals which we have met so far:

- The $1s$ and $2s$ AOs are spherically symmetrical and never change sign, so they are σ orbitals.
- The HAO on the lithium atom in LiH and the HAO on the nitrogen atom in NH_3 are σ orbitals.
- The MO occupied by two electrons forming the electron-pair bond in LiH is a σ orbital, as is the bonding MO in each of the N–H bonds in NH_3.
- The four $2p$ AOs of the two nitrogen atoms perpendicular to the bond in N_2 and the similar four $2p$ AOs of the carbon and oxygen atoms of CO are all π orbitals.

---

[10]This definition is provisional and will be made more precise later.

- If we extend the idea just a little we can call the lone-pair MO in the NH_3 a σ orbital because it is so similar to the σ MOs of the bonds in NH_3, even though there is no bond axis to view along in this type of case.

Notice that, in accordance with the general practice of mathematicians, the more 'advanced' a concept is, the more grand is the symbol given to it:

> A σ (pronounced 'sigma') is a Greek 's', reminding us that σ orbitals look like s AOs when viewed end-on. Similarly, a π (pronounced 'pie') is a Greek 'p', so that we can remember that π orbitals look like p orbitals from a certain point of view.

Thus, we say that a double bond (at least in the organic molecules we have looked at so far) is composed of two different environment-insensitive electronic substructures: a σ bond (described by a σ MO) and a π bond (described by a π MO). It is usual to use rather loose terminology and talk about the electrons involved in the σ bond as 'the σ electrons' and those in the π bond as 'the π electrons', even though all electrons are (of course!) identical.

As usual for contour diagrams, the sign change of an orbital is indicated by a change from full to broken lines for the contours.

7.5   The σ and π Orbitals in C_2H_4 and CH_2O

Here are contour diagrams of some of the orbitals involved in the description of the electronic structures in the ethene and methanal molecules: HAOs, AOs, σ bonds, π bonds and lone pairs (of methanal). Since both of these molecules are planar, some straightforward terminology can be used in describing the diagrams:

- 'From above' means looking down on the plane of the molecule; in this view all the nuclei can be seen marked on the diagrams.
- 'From the side' means looking in the plane of the molecule, seeing only the two 'heavy' nuclei (C and O).
- 'From the end' means looking in the plane of the molecule at the midpoint of the C–C bond, i.e. no nuclei are visible.

7.5.1   *Ethene contours*

Some representative contour diagrams are given below:

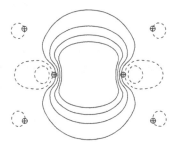

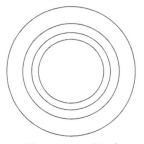

The C–C σ bond of ethene
viewed 'from above'

The same orbital
Viewed 'from the end'

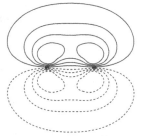

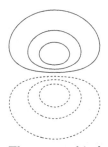

The C–C π bond of ethene
'from the side' of the molecule

The same orbital
viewed 'from the end'

The two right-hand contour diagrams illustrate very graphically the distinction between the electron distribution of a σ (Greek s) bond and that of a π (Greek p) bond.

The C–H bond (σ) MO has the appearance characteristic of C–H bonds in many organic molecules:

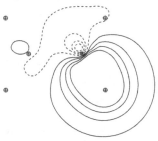

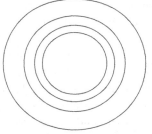

One C–H bond (σ)
viewed 'from above'

The same bond viewed
along its length

The other C–H bond is exactly the same but points at the other hydrogen atom.

To emphasise the appearance of the two important σ bonds, here are the two MOs (C–C and C–H) in perspective, slightly tilted to get a view of the appearance of the contour close to the nucleus:

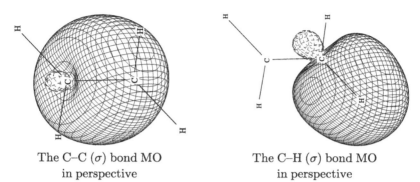

The C–C (σ) bond MO   The C–H (σ) bond MO
in perspective   in perspective

Looking carefully at these diagrams enables one to appreciate something of the nature of the electron distributions involved. Remember that the (3D) perspective views are only of the *outer* contour of the (2D) contour diagrams given on page 134. These perspective views of just one contour enable the whole shape of the distribution to be seen 'at a glance' rather than having to use the two diagrams of, for example, page 134, but they do not give the same information about the way the distribution changes with, say, distance along the bond the way the two 2D contour diagrams do.

7.5.2 *Methanal contours*

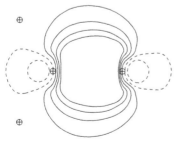

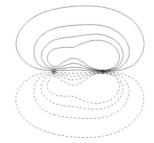

The C–O σ bond of methanal   The C–O π bond of methanal
viewed 'from above'   viewed 'from the side'

The views of the σ and π bonds 'from the end' are omitted, since they are almost identical to those of ethene given above. Notice that the asymmetry of the heteronuclear π bond is easier to see than that of the σ bond.

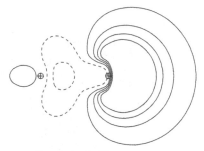

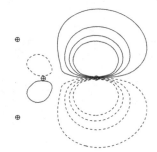

One 'σ-type' lone pair on the O atom, viewed 'from above'

The other 'π-type' O atom lone pair, viewed 'from above'

The strict σ/π terminology has been misused in the captions to the above figures, since clearly both lone-pair orbitals do not change sign when reflected in the molecular plane, and they are not involved in a bond, so there is no internuclear axis to look along. Nevertheless, it is clear that one of these lone pairs looks rather like the σ lone pair in (e.g.) NH_3 while the other looks like a $2p$ AO.

Organic tutors may be surprised to see these last two diagrams; they are used to claiming that the two lone pairs in methanal are *equivalent* in the sense that they are both exactly the same apart from orientation in space, in much the same way that the two C–H bonds are equivalent. For these tutors, here is such a pair of equivalent lone pairs.

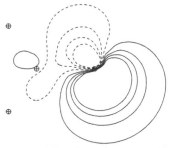

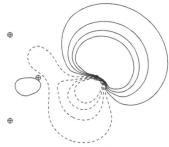

One 'sp^2' lone pair on the O atom, viewed 'from above'

The other 'sp^2' O atom lone pair, viewed 'from above'

The reason for this apparent ability to make a subjective choice of the way in which these lone pairs are presented involves a knowledge of the properties of quantum theory.[11] For the moment, it is enough to say that the *total* electron density due to these four electrons (two lone pairs) is exactly the same, whichever one of the two sets of orbitals they occupy. The difference lies in how we decide to divide this total density up into two separate pairs: where we 'draw the lines' around areas of space which contain any two electrons.

The C–H (σ) bond of methanal shown in the same two views as the C–H bond of ethene shows the similarity of the two molecular electronic substructures:

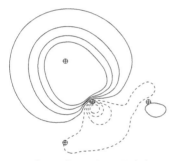

One C–H bond (σ)
viewed 'from above'

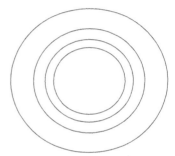

The same bond viewed
along its length

Comparison of these two diagrams with the corresponding ones for the ethene molecule on page 134 provides (at last!) some theoretical support for the assertion that the C–H bond actually *is* insensitive to its environment, which has been the main assumption taken from the data of empirical chemistry.

Although methanal is the simplest aldehyde and so the simplest organic molecule containing a C=O double bond, it is very typical of this commonly-occurring functional group. We shall want to refer to these bond and lone-pair orbitals again when we look at the electronic structure of the oxygen molecule in Chapter 9.

---

[11] Which we will address in Chapter 8.

7.5.3   *Relative energies of the two bonds*

In what was said above the σ bond was described, without much justification, as the low-energy bond of the two making up a double bond. This can be interpreted in two ways, depending on the area of interest.

(1) It can mean that the two electrons involved in this bond hold the two bonded atoms more tightly than those of the π bond, thus lowering the energy of the molecule more than those of the π bond.
(2) It may mean that the electrons in the σ bond are held more tightly to the molecule (by the attraction of the atomic cores) than the π electrons.

These two interpretations are not the same since, for example, any electrons in atomic cores (the $1s$ cores of carbon atoms, for example) are held very tightly in the molecule but contribute nothing to the binding energy of any bond. The simplest of these numbers to calculate are energies required to remove the electrons from the molecule; the so-called ionisation energies. These relative energy levels are, as we shall see, important in looking at the reactivity of double bonds, and are given below for some of the MOs of ethene and methanal:

Some calculated σ and π MO energies for ethene and methanal

Molecule	ϵ_{C-H}	ϵ_{C-C} σ	ϵ_{C-C} π
C_2H_4	−0.704	−0.986	−0.373
CH_2O	−0.770	−1.392	−0.530

Energies in atomic units (a.u.): $1\,\text{a.u.} \approx 2600\,\text{kJ}\,\text{mol}^{-1}$

These energies are an indication of how tightly electrons in the MOs are held by the molecule, *not* their contribution to the bond energies. The electrons in both the C–H and the C–C σ-bonding MOs are much more tightly bound than those in the C–C π-bonding MOs.

7.6   Reactivity of a Double Bond

We have seen in the previous subsection that the electrons comprising a π bond are much less tightly bound to the molecular framework than those of a σ bond. This means two things:

(1) They are relatively easy to remove from the molecule: their *ionisation energies* are smaller than those of 'σ electrons'.
(2) They are relatively easy to 'move around' in and near the molecular framework.

The second of these two points is all-important for chemistry:

> The distribution of π electrons in a multiple (double or triple) bond can be easily distorted by the approach of another molecule. This is what makes unsaturated molecules much more reactive than saturated ones.[12]

We can take another point of view of this fact. We have seen already that lone pairs of electrons have a distribution which 'sticks out' from the molecular framework — these electrons are repelled away from the bonds — and, in a less spectacular way, so do the π electrons in a multiple bond. So, just as lone pairs attract (and are attracted to) regions of positive charge and 'stick' to such regions via dative bonds, we might expect molecules with π electrons to be attracted to positive charges and form dative bonds. This does indeed happen and forms interesting species of organometallic compounds in which, for example, ethene attaches itself 'sideways' to metal atoms.

The upshot of these simple arguments is that the presence of multiple bonds in a molecule gives it a complex and interesting chemistry which is not present in saturated molecules. Although perhaps not quite so important in chemistry, the fact that the π electrons are easy to move means that unsaturated molecules have excitation energies which are much lower than those of saturated systems: they are often *coloured*.

7.7 Multiple Bonds in General

The molecules which have been considered in discussing double or triple bonds have been either linear or planar and, clearly, it would be far too restrictive to limit the idea of multiple bonds to *just* such molecules. The σ and π classification of AOs and MOs obviously strictly only applies to linear

---

[12]This, after all, is what chemistry is: the reorganisation of electrons and the resulting effects on the molecular framework.

or planar molecules, but this idea is easily extended to the more general case by applying the σ/π scheme to the 'local symmetry' of a given bond.

So, we adopt the following general scheme:

> If the 'local environment' of a given pair of atoms can be thought of as linear or planar, i.e. the atoms joined to both of the pair of atoms are either in line with those of the pair or in the same plane, then the σ/π classification of the (H)AOs and MOs can be used to define any single or multiple bonds between those atoms.

This is nothing more than saying that double bonds can be described anywhere in *any* suitable molecule by exactly the same methods as have been used for ethene and methanal.

7.8 Assignment for Chapter 7

Molecules which are not symmetrical often have electron distributions which are not symmetrical[13] and so have a dipole moment; they have no overall charge but the total charge distribution (that is, the charges on the nuclei and the electron distribution) is polarised. It is obvious that the methanal molecule should be such a molecule since, although it has two planes of symmetry, it is not symmetrical when the CH_2 group and the O atom are interchanged — any dipole moment will be along the direction of the C=O bond.

A valence-bond simple calculation of the electronic structure of the methanal molecule (using the HAOs illustrated in this chapter) gives the following approximate result for the electron distribution of the C=O double bond:

Electron populations of HAOs

Bond Type	C HAO	O HAO
σ bond	0.67	1.33
π bond	0.85	1.15

---

[13]The opposite is not true since, strictly, a molecule which is not symmetrical because of isotopic substitution may well not have a measurable dipole, since the electron distribution is determined largely by the charges, not the masses of the nuclei.

The electron populations in the C–H σ bond are 1.17 for the C HAO and 0.83 for the H HAO and, of course, the populations of the two atomic cores and the O lone pairs are 2.0 in each case.

Look at the diagrams of the HAOs in Section 7.5.2 on page 135, think about what the occupation of these HAOs means for the electron distribution, and comment on these numbers:

(1) Why is the σ bond stronger than the π bond?
(2) What do the numbers say about the shielding of the π bond by the σ bond?
(3) Do these numbers together mean that the number of electrons on the C atom is $2 + 0.67 + 0.85 + 2 \times 1.17 = 5.86$, the number of electrons on the O atom is $2 + 2 \times 2 + 1.33 + 1.15 = 8.48$ and 0.83 on each H atom? Notice that they do add up to the total number of electrons in the molecule.

In spite of the footnote on the previous page, can you think of a way in which the masses of the nuclei might affect the *measured* dipole moment of a polar molecule?

Chapter 8

Molecular Symmetry

Many of the simple molecules that we have discussed so far are symmetrical in the sense that, in the molecule, there are atoms which are in identical environments: the H atoms in CH_4, the H atoms in NH_3, the C atoms in C_2H_4, for example. In the ethene and methanal molecules of the last chapter, in both cases as well as the H atoms in the molecules having identical environments, the molecules are planar; this is another type of symmetry in the molecule (the molecule N_2H_4, for example, is not planar). Some accounts of the electronic structure of molecules make much use of elements of molecular symmetry. We need to make at least a short examination of symmetry in the molecular context.

Contents

8.1 The Question of Symmetry

We have already said that, for example, C–H bonds in organic molecules
tend to be very similar, having similar electronic structure and *very* similar bond energies (about 99 kJ mol^{-1}). Of course, when two C–H bonds
(or any other bonds) are in *identical* environments they must have *exactly*
the same electronic structure and bond energy. The question is:

> Is this fact of some fundamental importance and does it help
> in describing the electronic structure and energetics of sym-
> metrical molecules?

It is not at all difficult to see that the *total* electron density in any
molecule must reflect the forces acting on the electrons. In particular, the
major effect on the motions and distributions of the electrons is the strong
attractive force exerted on them by the atomic cores. The atomic cores are
always effectively spherically symmetrical around their nucleus. Thus the
total electron distribution must have the same symmetry as the nuclear
framework of the molecule. We have already seen this explicitly in the
case of the hydrogen molecule, where the atomic cores *are* the nuclei: the
electron distribution in the hydrogen molecule is identical on both sides of
a plane drawn through the midpoint of the two nuclei and perpendicular
to the bond direction.

Of course, when the quantum equations are solved and the electron distribution for a molecule of some given geometry is calculated, this electron
distribution does, indeed, have the same symmetry as the nuclear framework. So, strictly, one does not have to worry about the symmetry of the
distribution; it 'comes out in the wash'.[1]

So much is reasonable, perhaps even obvious. What is not so obvious
is the answer to the question:

> Does the distribution of each individual electron (or each
> electron pair) have to have the symmetry of the nuclear
> framework?

In an earlier chapter some prominence has been given to the existence and
properties of *lone pairs* of electrons, and, obviously, a lone pair is essentially associated with a single atom. So, if a molecule has some elements

---

[1]On the occasions where this is not true, it is always because some invalid approximations have been made in carrying out the solution of the equations.

of symmetry[2] which dictate that two nuclei are equivalent 'by symmetry' then a lone pair of electrons associated with one of these atoms does not have the overall symmetry of the nuclear framework. For example, a lone pair on one of the nitrogen atoms of N_2 is obviously not symmetrical with respect to reflection in the bond-midpoint plane.

Similar considerations apply in the case of the methane molecule discussed earlier. No mention was made of the obvious fact that any one of the C–H bond orbitals does not have the very high symmetry of the methane molecule. What is equally obvious, however, is that if there is a lone pair on one of the nitrogen atoms in the N_2 molecule, there must be an identical lone pair on the other (symmetry-equivalent) nitrogen atom. Similarly, if there is a pair of electrons involved in one C–H bond in CH_4, there must be three more identical structures associated with the other equivalent C–H bonds.

Precisely analogous remarks apply to the electrons in the atomic cores. The atomic core of a nitrogen atom consists of the nitrogen nucleus *plus* a pair of electrons in an atomic $1s$ AO. Clearly the distribution of these two electrons does not have the symmetry of the molecular framework; they are concentrated on one of a pair of symmetry-equivalent nuclei.

It is now completely clear what has to happen:

> In a symmetrical molecule, electronic structures associated with individual bonds or atoms etc. do *not* have to have the overall symmetry of the nuclear framework. *But* where there are environments which are identical by symmetry, *identical molecular environments must have identical associated electronic substructures.* That is, electronic substructures of the overall molecular structure must be *permuted* by the symmetry operations of a molecule in exactly the same way that symmetry-equivalent nuclei and atomic cores are permuted by those symmetry operations.

- Reflection in the bond-midpoint plane of the N_2 molecule must permute the two lone pairs. Similarly, this reflection, which permutes the two nuclei, permutes the two atomic cores, i.e. permutes the two $1s$ inner shells.

---

[2]Mirror planes, rotation axes etc. which, when a symmetry operation is performed (reflection in a plane or rotation about an axis) leave the molecule looking identical.

- Any of the 24 symmetry elements of the tetrahedral CH_4 molecule must permute the equivalent C–H bond electronic substructures *exactly*.
- But, reflection in the bond-midpoint of the N_2 molecule of the electron-pair *bond* must give exactly the same distribution, because there is only one of these and it is distributed around two symmetry-equivalent atomic cores.

8.2 Symmetry: Generalisation

The above considerations are not splitting hairs. They are vital to developing a theory of molecular electronic structure which reflects the actual environment-insensitive substructures in a molecule. Consult any book of data on chemical bonds and you will find that the C–H bond length for a single C–H bond, its bond energy, vibration frequency etc., are, to a very good approximation, independent of the environment it finds itself in; the C–H bond is *very* similar, not only in CH_4, C_2H_6, ... C_nH_{2n+2}, but also in any alcohol or ester, in morphine or cholesterol.

> If we try to insist that each *individual* electron pair has a distribution which strictly reflects the spatial symmetry of the nuclear framework of a molecule, we fly in the face of the most solidly-established fact of modern chemistry: the electronic structure of molecules is largely composed of environment-insensitive substructures.

The application of this strange 'rule' — that the distribution of each electron must have the symmetry of the nuclear framework — is most likely to be encountered in earlier accounts of the electronic structure of transition-metal-containing molecules — the 'complexes' of inorganic chemistry — largely because of the history of this subject being bound-up with spectroscopic, rather than chemical, considerations. Thus, one might see an account of the bonding in $Cr(CO)_6$ in which all the orbitals have the symmetry of the regular octahedral nuclear framework. In such an account there will be no orbitals which describe, for example, an individual Cr–C electron-pair bond or the CO bonding scheme. There will, however, be orbitals describing the lone pairs on the central Cr atom because it is a unique atom. The electronic spectrum of this molecule is due to transitions among these lone pairs which are characteristic of the (unique) central atom. The description of the UV spectrum of this molecule is the same, whether the symmetry 'rule' is applied or not.

If one were to study the structure of (e.g.) $Cr(CO)_4(N_2)_2$, with its much lower symmetry, one would find a very different set of orbitals bearing very little resemblance to those of $Cr(CO)_6$, even though the Cr–C bond is very similar in the two cases.[3]

8.3 Case Studies: H_2O and Benzene

The points being made here can be pressed home by a couple of examples.

Just one important introductory point: strictly speaking, it is the electron *distribution* which we are concerned with here, rather than the orbitals. The electron distributions are the *squares* of the orbitals, so that the requirement that a distribution is unchanged (or permuted with another) includes the possibility of a change of sign in the orbital. Thus, an *orbital* may be sent into plus or minus another orbital (including itself) without changing the requirement that the distribution is unchanged (or permuted).

8.3.1 *The H_2O molecule*

The water molecule is small enough to be able to make both a clear distinction between MOs which each have the symmetry of the molecule and MOs which describe the local electronic structure and are permuted by molecular symmetry operations.

It is quite possible to ignore all the qualitative 'chemical' considerations we have been using throughout, and simply use the 'brute force' method to calculate the MOs of any molecule; in particular one can use just *one* of the ideas we have stressed:

> In any region of space close to a charged nucleus the distribution of electrons will be rather similar to the distribution in an atom; that is, one can use AOs as building blocks to form MOs and then rely on the application of quantum theory to sort everything out.

---

[3]This comparison is shown explicitly in Chapter 16.

There are technical, computational reasons why the symmetry-adapted MOs are generated by this method.

If this simple project is carried out for the water molecule, one can obtain the MOs, each of which is adapted to the symmetry of the molecule, in the sense that the distribution of each pair of electrons[4] has the symmetry of the molecule. The two doubly-occupied MOs obtained in this way which involve electron density in the O–H bond region are given — as usual by contour diagrams — below.

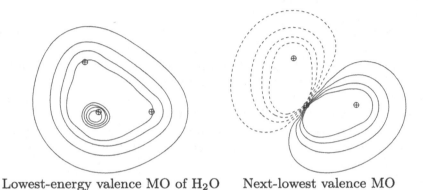

Lowest-energy valence MO of H_2O Next-lowest valence MO

In contrast, here are the two (equivalent) O–H localised MOs of the type we have been using thus far:

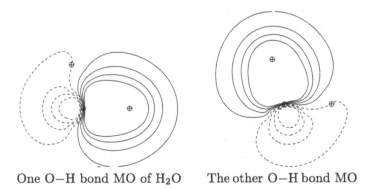

One O–H bond MO of H_2O The other O–H bond MO

---

[4]The electron distributions are the *squares* of the MOs.

The question naturally arises:

> What, if anything, is the connection between these two pairs
> of MOs?

The answer was hinted at in the discussion of the two possible representations of the lone pairs of the oxygen atom in the methanal molecule in Section 7.5.2

> When the total *electron density* due to the four electrons in
> the symmetry-adapted pair of MOs is calculated and compared to the total density due to four electrons in the two
> O–H bond MOs, the two are found to be *exactly* the same.

This fact has some interesting consequences:

- From a certain point of view one could say that it is immaterial which MOs one uses to describe the electronic structure of H_2O, since these two sets both give the same overall description of the total electron distribution.
- This same point of view would encourage the opinion that all the stress which has been placed on the *interpretation* of the overall electron distribution into bonds, lone pairs and the rest is just so much fantasy: there are no 'bonding pairs' or 'lone pairs' of electrons; where is the *evidence* for dividing the electron density into environmentally-insensitive substructures?
- How does this result fit in with experimental chemistry?

What is certainly true is that, if one has no knowledge of the science of chemistry over the past century or so, one might think that the use of bond- and lone-pair MOs was indeed arbitrary. If one were only interested in the total electron density of molecules, the way in which one partitions that density into electron pairs would be of no concern at all. However, if one were to consult the many shelf-miles of chemistry books and journals in university libraries, one would be quickly convinced that the evidence is *overwhelming* that the electrons in molecules do indeed behave in ways that suggest the division of the electron distribution into the chemical substructures which we have been using.

We can generalise from the above four-electron case:

> We can take the total electron density due to any number
> of electrons — a molecule, say — and divide that total up

in *any* way we like: into individual electron distributions, into electron pairs, into ones and threes etc. providing these components add up to the given total. But there are only a few — one usually — subdivisions which *nature* chooses: the one(s) which chemists use to elucidate the properties of that electron distribution.

So, the choice is between the following two possibilities:

(1) We *start from* the idea — based on the science of chemistry — that the electronic structure of molecules is composed of environment-insensitive substructures and use Coulomb's law, Schrödinger's mechanics and the Pauli principle to elucidate those structures.

(2) We simply solve the quantum mechanical equations and then look for some evidence in the results of our calculations that bonds, lone pairs etc. actually do exist.

In the latter case an A–B single bond (for example) would be 'discovered' by intentionally superposing the electron distributions due to at least two and perhaps many more of the symmetry-adapted MOs of the molecule.[5] This choice was made in Chapter 1 when we considered the question 'what are molecules made of?'

Consider if, perhaps on some other planet, quantum mechanics had been discovered *before* the familiar (to us!) simple rules of chemistry had been found. Suppose that it were possible to calculate the electron density of (what we call) molecules with great accuracy by using quantum theory. This alien civilisation would have found many, many different electron distributions for molecules and simply had to record them as if they were just facts. Imagine the reaction if some lonely chemical worker had proposed that the main features of all these varied calculations could be explained by a few (to us) simple ideas, the main one being the existence of environment-insensitive electronic substructures. Such a person would be celebrated as a genius. He (she or it) would be given the alien equivalent of a Nobel prize and their name would go down in alien history as the founder of a completely new science.

---

[5] A valid comparison is between regarding the solar system as simply a system of a couple of dozen bodies and solving Newton's equations *without using the fact that this solar system is composed of planets moving round the sun taking their moons with them as they go.* That is, simply ignoring the actual overall structure of the system and being compelled to find this out by looking at the complex path of each planet or moon.

8.3.2   *The benzene σ system*

The π-bonding scheme in the benzene molecule will be discussed in Chapter 11, but the σ-bonding scheme presents no difficulties using the ideas developed so far:

- The six carbon atoms are held together in a hexagonal ring by 'standard' σ bonds between pairs of carbon atoms qualitatively similar to those between the two carbon atoms in ethene.
- The hydrogen atoms are joined — one to each carbon atom — by C–H bonds similar to those in ethene and methanal.

But the six carbon atoms of the benzene molecule form, in fact, a *regular* (planar) hexagon; a highly symmetric shape. If, therefore, we perform a molecular orbital calculation in which each electron-pair occupies an MO which is constrained to have the symmetry of the molecule, then we generate a set of 12 σ-type MOs. These are the MOs occupied by what chemists consider to be the 12 electrons in the C–C bonds, and another in the six C–H bonds. Each of these MOs, when squared, gives an electron distribution which reflects the symmetry of the benzene molecule. So *each* of these 12 MOs may have, for example, contributions of equal magnitude from *all six* AOs (or HAOs) which are equivalent by symmetry on the carbon atoms and from the equivalent hydrogen AOs. Here is a representative selection of these MOs:[6]

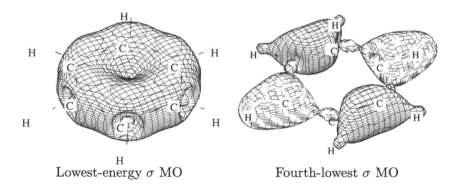

Lowest-energy σ MO   Fourth-lowest σ MO

---

[6]These contours are the (unsquared) MOs, with the broken lines being the negative lobes.

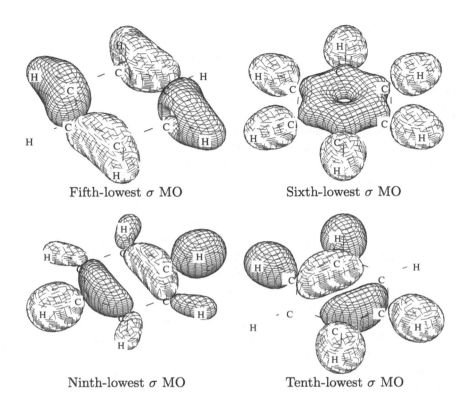

Fifth-lowest σ MO   Sixth-lowest σ MO

Ninth-lowest σ MO   Tenth-lowest σ MO

It is even more obvious here than in the case of the H_2O molecule that there is no hint in the form of these MOs that:

- The chemistry of the σ-bond framework of the benzene molecule is perfectly well described by a system of localised bonds between adjacent carbon atoms and adjacent carbon and hydrogen atom pairs.
- These localised bonds can be described by localised MOs: a set of six equivalent C–C MOs and a set of six C–H localised MOs.

When one carries through a calculation using the chemically-obvious bonding scheme, one finds the expected result: there are indeed bond MOs reflecting the qualitative picture of the σ-bond framework. Here are one each of the six equivalent C–C and C–H *bond* MOs:

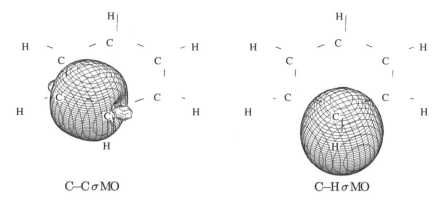

C–C σ MO   C–H σ MO

The symmetry of the molecule is, of course, reflected in these MOs just as it was in the delocalised set of MOs, but in a different way:

> The molecular symmetry is shown in the two sets of six MOs (C–C and C–H) by the fact that symmetry-equivalent electronic substructures (bond MOs in this case) are *permuted* by the symmetry operations rather than *each* electron-pair reflecting the symmetry individually by being transformed into itself or minus itself.

8.4 Bond MOs and Symmetry MOs

From now on little separate consideration will be given to symmetry considerations because, after all, if we get the electronic structure of the bonds, lone pairs etc. right, the symmetry will take care of itself, since it is the *attractive potential* of the nuclei which generates *both* the molecular symmetry *and* the electronic structure of the molecule. To use symmetry as if it were an independent consideration over and above the potential due to the nuclei is to put the cart before the horse.

> This sounds reasonable enough. *But* it is sometimes misleading because, while it is true that a quantitative *calculation* of the electronic structure will generate a structure with the correct symmetry, we are principally concerned with a *qualitative* approach to the problem of molecular structure, and, in spite of producing convincing models of the substructures,

we will have to come to terms with the fact that, however clever and convincing any scheme seems to *us*, nature will continue to outwit us and have new structures that we have not (yet) considered.

So, there can be no question of setting up a simple scheme and applying it unthinkingly in all cases; at every stage we have to be alert to new possibilities. This is true even when we look at the apparently simple case of diatomic molecules. More surprisingly, when we come to look at the structure of the π-bonding in benzene in Chapter 11, we shall find the opposite of our findings in the previous section: the electronic structures *really are* delocalised and the set of delocalised orbitals *does* give a good physical picture of the actual electronic substructure!

The most important idea to take away from these brief considerations of molecular symmetry in the context of molecular electronic structure is:

> If a calculation of the MOs of any molecule is carried out *without regard for the chemical structure of the molecule*, the MOs thus calculated do not, in general, give any indication of the *qualitative* aspects of the electronic structure. In particular, the forms of the resulting MOs cannot tell us whether the electronic structure of the molecule consists of localised or delocalised structures.

This is most obvious in the case of the benzene molecule. Straightforward calculations of the MOs of benzene produce MOs which are all (σ and π) adapted to the symmetry of the nuclear framework. But the *sigma* framework of the molecule is composed of 12 (six C–C and six C–H) strongly localised bonds, while the π electron structure is the model for the other delocalised structures of organic chemistry.

8.5   A Cautionary Note

These rather dismissive comments on the importance of symmetry only apply to the central problem which has been addressed here: the theory of the environment-insensitive electronic substructures in molecules. If we want to have a theory of molecular electronic structure more general than this, we can not necessarily carry over all of these ideas.

What has been said so far really only applies to the electronic structure of *isolated* molecules: nothing about collisions, chemical reactions or other main disturbances which molecules are prone to undergo. For example, if we want to look at what happens to molecules when we seriously disturb their electronic structure, we might have to think again about the applicability of molecular symmetry. Hitting a molecule hard with electromagnetic radiation is just such a case: for certain wavelengths of such radiation the electrons in molecules may take up entirely new distributions (molecular spectra in the visible and ultra-violet regions), or electrons may be knocked out of the molecule altogether (ionisation). In the case of ionisation, if an electron is removed from the molecule it has to be removed from one of the electronic substructures and, when this happens, the remaining electronic structure may well not have the symmetry of the molecule, so that substantial reorganisation of the electron density will have to occur. Excitation or ionisation of a molecule will usually be accompanied by changes in molecular geometry and corresponding changes in electron distribution.

Similarly, if a molecule is involved in a chemical reaction, the initial approach of the reacting partner will both destroy the symmetry of the molecule and distort the electron distribution. And, of course, any actual reaction involves considerable changes in electron density: breaking of bonds, formation of new bonds and all the effects on the rest of the two molecules transmitted by electron repulsion.

These types of change to the environments in molecules may require some considerations quite different from the ones we have been using. This is particularly true in the case of ionisation, which involves large changes in the electron distribution and, consequently, changes in the bond lengths and overall molecular geometry.

8.6 Assignment for Chapter 8

You will find that most texts on quantum chemistry or physical chemistry text-books present a different view on the use of symmetry in the theory of molecular electronic structure to the one taken in this chapter; symmetry is given quite a prominent treatment.

(1) Have a look at a few physical chemistry or quantum chemistry books and compare the two approaches, then discuss your opinions with your colleagues and tutors.

(2) Calculations of the electron distributions of molecules using the molecular orbital method can be simplified by using the idea that each *orbital* has the symmetry of the nuclear framework. This explains why one often finds the electronic structure of molecules containing only σ bonds given in terms of delocalised MOs. Do you find this *chemically* satisfactory?

Appendix F

Buridan's Ass and Molecular Symmetry

In fourteenth-century Paris, John Buridan was an eminent philosopher and theologian keenly interested in the problem of the freedom of choice; do humans really have the ability to choose for themselves, or are all their actions predetermined by God? He had the misfortune to be mocked by the inventors of the 'Paradox of Buridan's Ass'. How would a hungry donkey make a choice if it were placed exactly midway between two identical piles of hay; was this a free choice or ordained by God; would the donkey be paralysed by indecision and starve to death?

Some quantum theorists have suggested that this situation is similar to a supposed symmetry-related paradox in the electronic structure of molecules. The simplest example is that of the simplest molecule, which is highly symmetrical: the one-electron hydrogen molecule cation H_2^+. The problem is this:

As we have noted, the probability distribution of the electron in H_2^+ must reflect the symmetry of the attractive potential due to the nuclei and, in H_2^+, this means that the probability distribution must always be the same on each side of a plane bisecting the internuclear axis — that is be the same in identical points near each nucleus. But, if the molecule is pulled apart and the internuclear distance becomes greater and greater, at some point the single electron must 'decide' to which of the two identical nuclei it will be attached, since the products of the dissociation of H_2^+ *must* be a hydrogen atom and a proton. Which hydrogen nucleus does the electron 'choose' since both are identical?

Quantum theory says that the probability distribution must be the same around each proton, but any real *individual*

H_2^+ molecule must actually become *either* a proton and a hydrogen atom *or* a hydrogen atom and a proton:

$$H_2^+ \longrightarrow p^+ + H$$

or

$$H_2^+ \longrightarrow H + p^+ \ ?$$

This 'problem' is easily solved if we think about the interpretation of probability distributions:

- First, think about what might happen if one were to carry out an experiment with a large number of donkeys placed in the unenviable position above. One might very well expect that, if the stubborn beasts would co-operate, there would be approximately the same number choosing to go to the pile of hay on the left as to the pile on the right. If one had done enough trials the results would be a reasonable statisical approximation to the *probabilities* of each possible choice for the baffled donkeys. So, the probability distribution for donkeys choosing between the two piles would be the same.
- But probability distributions are exactly what the Schrödinger's equation calculates. Although it is easy to *imagine* pulling an individual H_2^+ ion apart, it is, in fact, not possible to do this experimentally and the solutions of the Schrödinger's mechanics reflect this fact; they only provide *probabilities* which would have to be verified by performing many, many such dissociations. Of course, if such experiments were carried out, they would reveal the same situation as many identical experiments with hungry donkeys; namely that for every case where the electron stayed with the proton of the left, there would be a case where the opposite 'choice' was made.

Quantum theory does not provide information about any individual example of a molecule or molecular process any more than the result of any individual throw of a die can be obtained from the probability theory of dice-throws. Probabilities can only be verified in practice by using statistical methods performed on many identically-prepared experiments. In the case of molecules any measurement will always be on billions and billions of individuals.

In a few words, there is no paradox here because the *probabilities* calculated for the H_2^+ molecule dissociation will reflect the symmetry of the nuclei even though any *individual* H_2^+ must 'choose' which way to dissociate.

Finally, there is what looks like the opposite 'paradox' in the molecular orbital theory of the more familiar hydrogen molecule: two electrons surrounding two protons. The normal model used for this molecule at internuclear distances around the observed value is, as we have seen in Chapter 4, two electrons of opposite spin occupying a single molecular orbital. Using this model at very large internuclear distances is obviously inappropriate since this would imply the same sort of 'choice' for the two electrons:

$$H_2 \longrightarrow H^+ + H^-$$

or

$$H_2 \longrightarrow H^- + H^+ \ ?$$

But, this time, the dissociation *is* actually symmetrical:

$$H_2 \longrightarrow H + H.$$

So, one may find statements in the literature asserting that the molecular orbital model 'breaks down' at large internuclear distances. It is not the MO model which breaks down but the incorrect assumption that, in the MO model, both the electrons *must always* occupy the same orbital, which is obviously incorrect at large internuclear distances; the result will always be one electron in a $1s$ AO on each atom. The resolutions of these two similar-looking 'paradoxes' are thus quite different: the first is a confusion about the interpretation of probabilities and the second an incorrect approximate model for the electronic structure at large internuclear distances.

Chapter 9

Diatomics with Multiple Bonds

A discussion of the electronic structure of 'simple' diatomic molecules — those composed of the atoms H through F — has been put off for long enough by saying that their structure is not typical. This chapter considers the possible molecules of the type X_2, where X is one of Be, B, C, N and O, and a slight diversion into the structure of the CO molecule. Some comparisons are made with the considerations of Chapter 7, where multiple bonds in organic compounds were discussed, and Chapter 8, where some of the implications of molecular symmetry were introduced. There is an important lesson to be learned from this excursion outside the domain of 'typical' polyatomic molecules.

Contents

9.1   Motivation

Unfortunately, as we have already noted, there is no simple diatomic
molecule with a 'simple' double bond of the type we have discussed in
the ethene and methanal molecules in Section 7.5.3, where the bonding
structure is two electron pairs, one of σ type and the other of π type.
The reason for this fact, which is surprising at first sight, is that as we
have already seen, the *overall* electronic structure of any molecule must
have the same symmetry as the nuclear geometry. That is, in the case of a
diatomic molecule, the *total* electronic structure must be cylindrical, with
the bond as the axis of the cylinder. In terms of bonding, we shall see that
this means if there is one π bond in a diatomic molecule, there must be
another one identical except for orientation in space. So to make a start
we must actually look at molecules with *triple* bonds — one σ bond and
two π bonds — in order to study multiple bonds in closed-shell diatomics.

The most familiar homonuclear diatomic molecules containing a multiple bond are N_2 and O_2; the other two homonuclear diatomics from the
first row — H_2 and F_2 — have only a single bond between the atoms.
The oxygen molecule is rather special because its electronic structure is
not composed of closed shells of electrons. In fact, it contains *two* unpaired
electrons, which give it its characteristic properties. Although unusual, this
is what enables O_2 to be thought of as having a double bond. The nitrogen
molecule *is* a closed-shell electronic structure, and so we begin the discussion, not from the beginning but from the middle of the periodic table row,
with the electronic structure of this common molecule.

9.2   The Nitrogen Molecule: N_2

Any diatomic molecule is necessarily linear and the approach of two atoms
will, as usual, polarise the atomic electron distributions to generate a pair of
HAOs on each atom, one concentrated in the region between the two atoms
and another 'pointing' in the opposite direction, away from the internuclear
region.[1] The two HAOs concentrated in the internuclear region are then
clearly suitable to be added together to form a σ-bond MO, and the other
HAOs are obvious candidates for lone pairs when each is occupied by two

---

[1] Why? Remind yourself of the way in which electrons on the same atom will repel
each other.

electrons. The basic 'σ framework' of the N_2 molecule then uses six of the ten valence electrons of the two nitrogen atoms (remember, there are four in the two $1s$ atomic core AOs). Here are contour diagrams of the three doubly-occupied σ MOs:

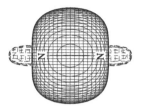

The σ-bonding MO of the N_2 molecule

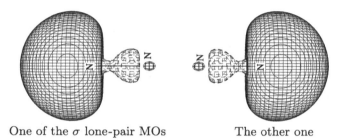

One of the σ lone-pair MOs   The other one

By a simple extension of the ideas used in the discussion of the ethylene (C_2H_4) molecule in Section 7.4, we can safely assume that the remaining four valence electrons of the molecule occupy two π MOs, each described by the combination of a pair of π AOs, one from each N atom. In summary, the overall electronic structure of the nitrogen molecule, is composed of seven electron-pair substructures:

- Two pairs of electrons: one pair in the $1s$ atomic core AO of each nitrogen atom.
- One pair of electrons: the σ bond described by a σ-bond MO formed from one HAO on each of the atoms.
- Two pairs of electrons: a lone pair on each nitrogen atom whose distribution is described by an MO formed mostly from a single HAO pointing away from the bond region.
- Two pairs of electrons: two π bonds, each described by a π-bond MO formed by the combination of one π AO ($2p$-type) from each atom. The relative orientations in space of the two π bonds will have to be determined.

The only new idea here is an important one: the idea of the *relative* distributions of the electrons in a situation where there are two π bonds between two adjacent atoms. Each π bond will have the characteristic appearance similar to the one in ethylene. But what is *overall* electron distribution of these four electrons? We have already seen that, once the bond structure in a molecule is formed, the overall orientation of the electron pairs and nuclei is determined by their mutual repulsions — by Coulomb's law — and this must be the case here as well. It is fairly easy to apply ideas of electron-pair repulsion to these substructures, since there is no complication due to the relative orientation of the nuclei: the molecule must be linear.

- In the absence of any π electrons, the three σ electron pairs would be linear. Their mutual repulsions would be minimised by the two lone pairs pointing away from their parent nucleus in line with the bonding electron pair between the nuclei.
- If we now introduce just one occupied π MO the electrons will, of course, repel the σ electrons, but this repulsion will not affect the shape of the molecule because it must be linear.
- The same remarks apply to a second pair of π electrons. The only question is how these two π-electron pairs mutually orient *each other*.
- The answer is obvious, simply by looking at a diagram. The π electrons must get as far away from each other as they can, and they do this by taking up positions 'at right angles' to each other, if one can describe the orientation of two distributions by such a simple term.

The diagram below makes the meaning of the statement much clearer than more words:

 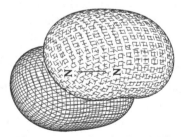

One of the π MOs of N_2 The 'perpendicular' π MO

As in the σ-MO diagrams above, this diagram is a 3D view of a typical *one* of the contours of each MO, but has been tilted slightly to give a clearer view of both. The electron distributions are the squares of the MOs and have an almost identical appearance. Notice what 'at right angles' means here: the planes bisecting each MO containing the bond axis are perpendicular in the same way as the $2p$ AOs are 'at right angles' to each other.

9.2.1 *Energies of the N_2 MOs*

It was mentioned at the start of Chapter 7 that the energy of the triple bonds in both N_2 and CO are anomalously high and very untypical of the multiple bonds in polyatomic molecules. The reasons for this are not clear and this question will not be addressed here. What *is* useful and relatively simple to do is to look at the relative energies of electrons in the various MOs of these molecules. The $1s$ electrons in the atomic cores are excluded from this discussion; they are essentially associated with the parent nucleus and very tightly bound to it.

The standard way to display this kind of information is by using an energy-level diagram. Such diagrams are always drawn with the most tightly-bound electrons at the bottom and less tightly-bound ones higher up. The gaps between the levels are an indication of the *relative* tightness with which the electrons are bound to the molecule by the attraction to the atomic cores.[2] In the diagrams the relationships amongst the various orbitals are shown by using dashed lines to indicate which AOs are used to form HAOs and which HAOs are the main contributors to the final MOs. In the case of N_2 we are concerned with the electrons originally in the $2s^2 2p^3$ structure of the constituent atoms. The idealised steps in molecule formation which have been used throughout can have their energies compared as follows:

- The $2s$ electrons are more tightly bound to the atom than the $2p$ electrons and, because the atom is spherically symmetrical, all three $2p$ electrons must have the same energy, since their distributions in space are identical apart from orientation.

---

[2]These simple rules are exactly the same as the rules used to sketch the *relative* energy levels available to energies in atoms used in Chapter 2.

- The electron distributions of the isolated atoms are polarised by approach to form distributions characterised by HAOs. These have been approximated by 'mixing' the $2s$ and $2p_\sigma$ AOs and, since a $2s$ electron lies deeper than a $2p$ electron, this will generate a pair of HAOs with energies between the two AO energies.[3]
- The two HAOs concentrated in the bond region are mixed to form a σ MO.
- The two sets of two π-type $2p$ AOs mix to form two π MOs.

These energy changes can be conveniently summarised on a qualitative diagram which simply indicates the *relative* energies of the various electrons in the orbitals. In constructing this diagram, the ideas introduced in Chapter 7 about the relative energies of σ and π MOs have been used, even though the contribution to the bond energy of the *two* π bonds appears to be greater than that of the σ bond. We always have to bear in mind that the tightness of the binding of the *electrons* to the nuclei does not necessarily correlate with the energy with which the nuclei are *bound together* by those electrons: an extreme example, of course, is provided by the inner shells of electrons in a molecule.

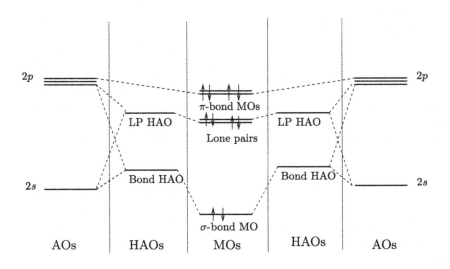

---

[3] Convince yourself that this is true.

Points to notice:

(1) Because of the molecular symmetry of a homonuclear diatomic molecule, the energies of the equivalent AOs, equivalent HAOs and equivalent MOs (π MOs and lone pairs) are identical.

(2) There is evidence from other molecules that the energy of the σ-bond MO is lower than the energy of the π-bond MOs.

(3) The energy of the σ-bond MO is expected to be lower than the energy of the lone-pair MOs. After all, they are all of σ type and the electrons in the bond MO are attracted to two atomic cores, while those in the lone pair are concentrated near one atomic core.

(4) What is not at all clear — because we have not needed to discuss it yet — is the relative energies of the π-bond MO and the lone pairs. Both the π-bond MO and the lone pairs are higher in energy than the σ-bond MO, but how do their energies compare *with each other*? In fact, detailed calculations show that, for N_2 at least, the σ lone pairs are lower in energy than the π-bond MOs, and this is shown on the diagram.

In the case of the nitrogen molecule, the relative energies of these MOs do not make any difference to our qualitative picture of the electronic structure of the molecule, simply because all three of them are doubly occupied, and our picture would be unchanged if the relative energies of the σ lone pairs and the π-bond MOs were different. With five valence MOs and ten electrons, our rules for filling up the orbitals leave no room for choice, whatever their energy order. But, as we shall see, this is a very important point when we look at other diatomic molecules with fewer than ten valence electrons,[4] where the situation is not so clear.

The nitrogen molecule is quite unreactive, but, from what has been said about the chemistry of multiple-bond-containing organic molecules, it might be expected that what reactivity the molecule has would be due to the π electrons. In fact, it turns out that the lone pairs of N_2 are the ones involved: N_2 forms dative bonds to some metals in which the $M-N\equiv N$ entity is (nearly) linear. Again, this points to the possibility that the energies of the lone-pair electrons and the π electrons might be comparable.

---

[4]Remember, any *homonuclear* diatomic must have an even number of electrons!

9.2.2 *Symmetry and the N_2 molecule*

The σ system of N_2 provides a convenient example to illustrate the ideas of symmetry outlined in Chapter 8. If one uses standard computational software — which uses the symmetry of the molecule to minimise use of resources — the σ MOs obtained each have the symmetry of the molecule. In particular all four occupied σ MOs are identical when reflected in a plane through the bond centre (apart, possibly, from sign). Here they are in 'slice contour' representation:

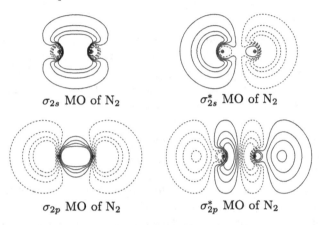

σ_{2s} MO of N_2 σ_{2s}^* MO of N_2

σ_{2p} MO of N_2 σ_{2p}^* MO of N_2

Clearly:

- The lowest MO — σ_{2s} — is very similar to the MO formed from the two HAOs in Section 9.2. Double occupancy of this MO forms the (electron-pair) σ bond.
- *Both* of the other two occupied MOs have more density *outside* the bond region than within it:[5] MO σ_{2s}^* has a large positive lobe on the left and a symmetrically equivalent negative lobe on the right, while MO σ_{2p} has a negative lobe on each end.

If we form the two new MOs:

$$\ell_1 = \sigma_{2p} + \sigma_{2s}^*$$
$$\ell_2 = \sigma_{2p} - \sigma_{2s}^*,$$

---

[5] Remember that the electron density is given by the *square* of the MO, so the negative (dashed) contours become positive contours of electron density.

we see that, in the case of MO ℓ_1, the left-hand lobe of σ_{2s}^* cancels the left-hand lobe of σ_{2p}, while, in the case of MO ℓ_2, the two left-hand lobes reinforce each other. Thus ℓ_1 and ℓ_2 are nothing more than a *lone pair* on each nitrogen atom pointing away from the bond region, exactly as we saw on page 163 of Section 9.2.

There is no hint of the fact that what limited chemistry the N_2 has is provided by the lone pairs in the symmetry-adapted MOs.

The biologically and industrially important CO molecule has a very similar electronic structure to N_2, but, although it has very different chemical properties,[6] it is useful to summarise its electronic structure here.

9.3 The Carbon Monoxide Molecule: CO

The qualitative difference between the CO molecule and its isoelectronic partner N_2 is that it is *heteronuclear*: six of the 14 electrons are from the carbon atom, and the remaining eight from the oxygen atom. The description of the qualitative features of its electronic structure is basically identical: the seven electron pairs described in the last section are the essential components.

What will be different, of course, are the details of the electron distributions due to the different nature of the atoms. In particular, the carbon atom core will attract electrons less strongly than a nitrogen core, while an oxygen atom core will tend to exert a stronger pull on electrons than the nitrogen core. This means that we might expect the CO molecule to have a dipole moment due to its asymmetric electron distribution. Also, the energy-level diagram for the MOs will not be symmetrical: the energies of the O-atom AOs are lower than those of the C atom, and so the lone pair on the carbon atom will be more loosely bound (therefore more chemically reactive?) than either that of the oxygen atom of CO *or* the lone pair on the N atom of N_2. Many of the main features of the diagram are similar, but the diagram is skewed with respect to the symmetrical N_2 diagram, and the lone pairs are no longer degenerate, being largely the same as the HAOs of the parent atoms. In the diagram below, the σ-system levels are given.

---

[6] Think of the difference between inhaling nitrogen, which you do thousands of times every day, and inhaling carbon monoxide, which you might only do a few times!

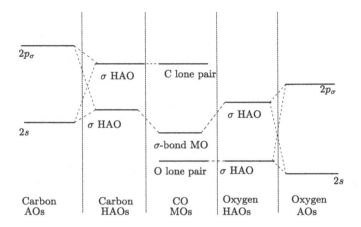

The C≡O bond is *very* strong; one of the strongest bonds in chemistry, and yet the molecule is extremely reactive. Unlike N_2:

- CO burns very readily. In fact, before the discovery of large deposits of methane it was the major source of domestic heating in the UK.
- It is very poisonous. It binds to haemoglobin more strongly than oxygen.
- It is an extremely important reagent in the chemistry of the transition metals, readily forming strong bonds of the type M−C≡O, in which the metal–carbon–oxygen entity is linear.

The chemical reactivity of the CO molecule is almost entirely due to the lone pair associated with the carbon atom; it retains its triple bond in most reactions. This is a notable contrast with multiple bonds between carbon atoms in organic molecules, where it is more typical for (e.g.) a double bond to react to add other atoms or groups to each of the doubly-bonded carbon atoms, and leave these two joined by just a single σ bond. Here are some calculated MO energies for these two molecules:

Calculated σ and π MO energies for N_2 and CO

Molecule	σ Bond	π Bond	Lone Pair 1	Lone Pair 2
N_2	−1.645	−0.620	−0.851	−0.851
CO	−1.622	−0.635	−0.678 (Carbon)	−1.025 (Oxygen)

Energies in atomic units (a.u.): $1\,\text{a.u.} \approx 2600\,\text{kJ}\,\text{mol}^{-1}$

The electron pairs responsible for the characteristic chemistry of both CO and N_2 — lone pairs — are actually held more tightly than the π electrons. It seems that, to be involved in chemical reaction, it is just as important for electrons to be *in the right place* as to be not too tightly bound.

9.4 Other Homonuclear Diatomics

The other homonuclear diatomic molecules which might possibly contain multiple bonds would be rather more exotic species; except for O_2, not the sort of thing to be met every day, even in a chemistry laboratory. The possibilities are: Be_2, B_2, C_2 and O_2. What is immediately obvious about all of these systems is that only the elements N and O exist as N_2 and O_2, i.e. *gases*. All the other elements are *solids*. Beryllium exists under ordinary laboratory conditions as a *metal*: it conducts electricity and is composed not of diatomic molecules, but vast arrays of mutually-bonded atoms with their electrons free to move over the whole crystal. Carbon and boron are also solids under ordinary laboratory conditions, but, rather than being metallic, they exist in a variety of crystalline forms in which the carbon or boron atoms are joined by covalent bonds to *many* of their neighbours. Carbon exists as graphite or diamond[7] and boron as a B_{12} basic unit which is an *icosahedron*.[8] It is, therefore, obvious that there is little point in using the methods that have been developed so far for small molecules to get a description of the common forms of these elements as diatomic molecules. More importantly, although Be_2, B_2 and C_2 are known experimentally, they are transient species requiring highly specialised techniques to prepare them under conditions of high vacuum. They occur naturally only in space and, although C_2 has an important role to play in reactions in interstellar space, all of them are found to have extremely unusual electronic structures which do not fall into our categories of having environment-insensitive substructures. Typically, the calculation of the electronic structures of these very small molecules is described in the chemical literature as being 'challenging,' and none of them has a single MO-type structure which adequately summarises their electron configuration.

However, the oxygen molecule is a common and very familiar molecule and, if we use the energy-level diagram sketched for the N_2 molecule as a

---

[7]Plus, of course, all the more recently discovered fullerenes and their crystals.

[8]A regular icosahedron has 20 triangular faces and 12 vertices.

guide, it might be possible to get a general idea of the qualitative electronic structure of this important homonuclear diatomic.

9.4.1   *The oxygen molecule: O_2*

This molecule has twelve valence electrons and so from the very start it can be seen that the diagram is inadequate since it only shows five MOs and so can accommodate ten electrons. This is merely because the diagram was generated with N_2 in mind. It can be simply extended to include the two MOs formed from the linear combination of the two sets of π (H)AOs with a negative sign. There is, however, the question about whether such a MO — which, after all, does not satisfy the general requirement of enhancing the electron density *between* the nuclei — would exist. This can only be found out by direct calculation. In fact, calculations show that, although electrons in these orbitals actually *diminish* the strength of the bond between the two atoms, the electrons in these MOs are actually (weakly) bound[9] to the molecule by the combined attractions of the atomic cores and the repulsions of the other valence electrons. For this reason, these π MOs are known as *anti-bonding* MOs. Here is a slice diagram and a perspective contour of one of them:

The O_2 π anti-bonding MO

---

[9] It is important here to distinguish between the two quite different senses in which 'bond' and 'bound' are being used. It is perfectly possible for electrons to be held captive ('bound') to a molecule, while at the same time not contributing to the bonding of the atoms in that molecule or even, as in this case, actually weakening that bonding.

It looks rather like a 'spread out' $3d$ AO in form.

The full MO energy-level diagram, augmented by the 'anti-bonding' MOs is then:

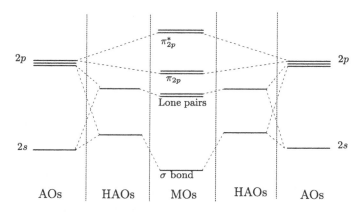

What *qualitative* picture of the molecule can be obtained from this detailed diagram? The two electrons in the highest MOs each occupy one of the degenerate pair of anti-bonding MOs for the familiar reasons giving the molecule two *unpaired electrons*. In summary:

- The σ system is qualitatively identical to that of N_2: one doubly-occupied σ-bond MO and two lone pairs. This generates one 'standard' electron-pair σ bond.
- There are four electrons in π-bonding MOs, which would provide two 'standard' electron-pair π bonds if it were not for the other two electrons in the π system.
- There are a total of two electrons occupying π-anti-bonding MOs, which actually partially counteract the bonding effect of the four electrons in the two bonding π MOs.

In order to 'count the bonds' in this molecule, strictly the combined effect of one 'anti-bonding' electron and two 'bonding' electrons in each of the two perpendicular π systems would have to be calculated and compared with a 'normal' π-bond energy.

In fact, to obtain a simple qualitative picture we simply count the *net number of bonding electrons* by subtracting the number of anti-bonding electrons from the number of bonding electrons. This gives us two 'half bonds', one in each

π system, so that the simple picture is that the O_2 molecule
has a double bond composed of a single σ bond and two π
half bonds,

and so, can be made to conform to the rules of mainstream chemistry in
its electronic structure.

9.5   Lessons from Diatomics

Perhaps the most important lesson to be learned from this simple look at
the electronic structures of the diatomics formed from the first row of the
periodic table is that, with the exception of N_2, O_2 and CO,[10] they do not
exist under normal laboratory conditions. Only these three molecules are
capable of exhibiting their 'normal' valencies as diatomic molecules, and
even CO has a carbon atom with a valence of three and a lone pair, instead
of its familiar valence of four. The other, unstable or transient species are
prevented from exhibiting their 'natural valencies' and take on extremely
unusual electronic structures, constrained as they are by the high symmetry
of their diatomic nature. The most spectacular example is C_2, for which
the nearest qualitative model is two C atoms joined by two π bonds and
no σ bond, which must be a unique situation, not just for the normally
four-valent carbon atom, but for any two bonded atoms.

The oxygen molecule, although in the main capable of being described
by the environment-insensitive substructures we have used throughout, has
a new type of π structure not met before. For symmetry reasons it cannot
form its 'normal' double bond consisting of a σ and a π bond. Instead, it
has to have a π structure of two doubly-occupied π-bond MOs — which,
apparently, would give the molecule a *triple* bond — whose bonding effect
is partially offset by two *singly*-occupied π MOs, weakening the π-bonding
and giving O_2 effectively its 'normal' valency of two.

If we had used the N_2 energy-level scheme to attempt to describe the
structure of the other, transient, molecules we would have come across
absurdities like a 'singly-occupied lone pair MO' which is a contradiction.
The overall message is clear enough:

> Any set of concepts and rules is prone to fail if it is used
> outside the area which it was developed to describe.

---

[10] And, of course, the simpler single-bonded ones, H_2, HF and F_2.

What has happened here is that, for these highly symmetrical small molecules, it does not seem to be possible to use our main idea of *transferable* environment-insensitive electronic substructures, because:

- The requirement that the overall electron density be the same symmetry as the nuclear framework seems to conflict with the idea of localised environment-insensitive structures.
- In particular, the idea of lone-pair orbitals was specifically introduced to explain:

  (1) The shapes and geometries of *polyatomic* molecules and,
  (2) That the basic assumption was that these lone-*pair* orbitals would be *doubly occupied.*

If we try to apply a model of electronic structure which was developed to describe closed-shell polyatomic molecules to exotic, transient molecules which contain unique substructures, we should not be surprised that it does not work! In Chapter 1 this danger was anticipated, when it was emphasised that to study a part of nature it is always necessary to decide on the 'large-scale' structure of the problem in advance of the use of more basic theories.

9.6 Assignment for Chapter 9

We saw in the previous chapter that the total electron distribution in a molecule must have the same symmetry as the nuclear framework. In particular the symmetry of a homonuclear diatomic molecule must have both the reflection symmetry which interchanges the two nuclei, and the symmetry of rotation about the bond axis. Of the molecules that we have met so far, diatomics have the highest symmetries, and homopolar diatomics have the highest symmetry of all. In the following problems ideas from the current chapter are used to illustrate molecular symmetry as well as the structure of diatomics.

Sketch the MO energy level diagram for the N_2 molecule on page 166 of Section 9.2 and think about what happens if the the molecule is ionised — one electron is removed. Obviously, since the molecule has ten valence electrons (ignoring the very tightly bound $1s$ electrons for the moment) there are ten ways of removing an electron but, because of the symmetry, not all of these are different. Remember how ionisation is indicated on an AO or MO energy-level diagram — it is the energy required to take an electron out of the system and is represented by a level *above* all the bound MO energies.

(1) Add the 'ionisation level' to your diagram.

(2) How many *distinct* ionisation energies are there in the molecule? Indicate them on your diagram by vertical arrows.

(3) How many of the ionisations which you found above will leave the molecular ion (N_2^+) with an electron distribution which has an electron distribution with the correct symmetry; i.e. an electron distribution with the same symmetry as the nuclear framework? You will need to look at the contour diagrams of the MOs in Section 9.2.

(4) Which MO(s) have the smallest ionisation energy — which electron(s) require the least energy to be removed? Presumably this will be the 'lowest ionisation energy'. Does this ionisation give a molecular ion with the correct symmetry?

(5) When one of the four most tightly-bound electrons in the N_2 molecule is removed — from one of the two doubly-occupied $1s$ AOs — it is found that the *first* detectable product formed is a molecule which has one singly-occupied AO on one of the N atoms, and one doubly-occupied MO on the other.

(6) Discuss the implications and interpretation of the last two items with your colleagues and your tutors.

Chapter 10

Dative Bonds

We saw at the end of Chapter 6 that lone pairs of electrons will be expected to undergo changes when given the chance to be attracted to positive ions or regions of positive potential. In this chapter we look at the consequences of this simple fact. We shall see that some of the simplest and most familiar chemical reactions — the 'everyday' ones, so familiar that we forget that they are chemical reactions — involve the formation of dative bonds.

Contents

10.1 Introduction: Familiar Reactions

It was noted at the end of Chapter 6 that the reorganisations of electrons associated with the formation of a molecule with covalent bonds leads, in some cases, to lone pairs. These lone pairs, because they are repelled by the electrons pairs in the bonds, will tend to be 'sticking out' of the molecule and will therefore tend to be attracted to sources of positive electrical potential:

- Positive ions (atomic or molecular).
- Regions near uncharged molecules which have regions where the positively-charged atomic cores are not as well shielded by electrons than the rest of the molecule.

This has some important consequences, not least that this process is exactly what happens in some *very* familiar chemical reactions. In fact, some of these reactions are so familiar that they are often not regarded as chemical reactions at all.

10.1.1 *'Solvation'*

Perhaps the most familiar cases of such reactions are those between the lone pair of a solvent molecule and a positive ion; for example, the mutual attraction between an ammonia (NH_3) or water (H_2O) lone pair and a hydrogen cation (a proton, H^+) to form, respectively, the ammonium cation (NH_4^+) or the H_3O^+ cation. What is happening here is a special case of the general principles outlined in Chapter 3:

- The (negatively-charged) lone pair of electrons is attracted by the field of the positively-charged proton and polarised by the attractive force of this field.
- There is no electron associated with the proton and so no mutual polarisation is possible here as there was in the earlier examples.
- As the two species approach, the electron distribution of the now polarised lone pair is further disrupted and the two electrons are shared between the nitrogen atomic core and the proton (the hydrogen atomic core), forming a covalent bond.

The difference between this sequence and the ones described earlier is, of course, that one of the partners in the act of bond formation does not contribute any electrons to the bond. Both are provided by the partner with the lone pair. However, the overall effect is the same:

> Two electrons have had their distribution completely changed by the coming together of two species and, as a result, the two electrons are *shared* between two atomic cores,

in the sense that the distribution of the two electrons is around and *between* the two atomic cores.

If we look at the structure of the NH_4^+ ion it is clear that the four N–H bonds are equivalent: the molecular ion is tetrahedral, with all four N–H bond distances the same. We cannot tell *after* the addition of a proton to NH_3 *which* of the four hydrogen nuclei were in the original molecule, and which one is the added proton. But, it is equally clear that one of the now-equivalent N–H bonds was *formed* by a completely different process from the other three. The differences are very obvious:

- The way to make ammonia from nitrogen and hydrogen is by way of the reaction between *molecular* hydrogen (H_2) and molecular nitrogen (N_2), and needs vigorous conditions (high temperature and a catalyst[1]).
- To make NH_4^+, all that is needed is to let ammonia get close enough to a molecule with a loosely-bound hydrogen atom and the ammonia will snatch the hydrogen atom. Just take the stoppers off bottles of ammonia solution and concentrated hydrochloric acid, hold the stoppers close together and watch the ammonium chloride form in the air!

Similar remarks apply to the formation of H_3O^+. The O–H bonds in H_2O are formed by burning hydrogen gas, while the other bond — although equivalent once formed — is formed simply by dissolving a molecule with a loosely-bound hydrogen atom — any acid — in water.

When the lone pair has captured a proton, how will the electrons re-arrange themselves in this new environment? We use the same arguments as we have done earlier, listed on page 68. In short:

- Close to each atomic core the distribution will be like that of the separate atom distribution (possibly polarised).
- Between the atomic cores the distribution can be accurately approximated by a linear combination of (H)AOs.

The argument is identical to that used in discussing the covalent bond: an MO is formed by a linear combination of (H)AOs. The only difference is that *both* the electrons come originally from just one of the atoms, rather than one from each. So, since the *result* is the same — all the four N–H bonds in NH4$^+$ are identical — one might argue that we should perhaps not distinguish strongly between the bond formation process of this

---

[1] Remember the Haber process?

type — historically distinguished by calling it a *dative* bond — and an
'ordinary' covalent bond. However, that is not at all to imply that there
is no such thing as *dative bonding*. From what we have just said, the *process* of forming a covalent bond by 'donation' of an electron pair, rather
than the sharing of one electron from each partner in a covalent bond, is a
distinct way of forming chemical bonds. Without searching for some new
and confusing name for this process, we can speak of a 'datively-formed'
covalent bond as the distinguishing feature.

The name 'dative' is a sensible one; the lone pair of a given molecule
donates a pair of electrons to the bond while, in a normal covalent bond,
each atom joined by the bond donates one electron to the bond. Notice
that one speaks of donation of electrons to the *bond* and emphatically not
donation to the bonded *atom*.

When attempting to give a qualitative description of *any* redistribution of electrons,[2] the question to ask is, as usual, 'how will the electrons
distribute themselves in a given environment' not 'how will atomic orbitals
combine?'

In this context one might ask:

> Why does one use the AO of the hydrogen atom as part of the
> expansion of the dative-bond MO when there is no occupied
> $1s$ AO on the hydrogen cation (the proton)?

This is a key point to be understood and is where the approach in this
book differs from the approach in most texts.

> AOs are not used to expand an MO because they are occu-
> pied in the constituent atoms of the molecule (although they
> often are), *but* because, in the region close to an atomic core,
> an electron *must* have a distribution close to that associated
> with an AO of that atom. Thus, when we wish to describe the
> electron distribution of the dative bond formed by the inter-
> action between a nitrogen lone pair and a hydrogen cation (a
> proton), this electron distribution must be similar to a nitro-
> gen (H)AO near the nitrogen atomic core — the lone pair
> orbital — and similar to a hydrogen AO (a $1s$ AO) — close

---

[2] And one can make a case for saying that this is what all chemistry is about: the
redistribution of electrons and its consequences.

to the hydrogen atomic core, *whether or not* either of these AOs are occupied in the separate fragments.

The appeal must always be to the laws of nature — electrostatics and Schrödinger's mechanics — not to rules-of-thumb involving manipulation of mathematical functions.

10.1.2 A reactive lone pair: the CO molecule

We have seen already that the diatomic carbon monoxide molecule has a lone pair of electrons 'sticking out' from each end; one from the carbon atom and one from the oxygen atom. The one associated with the carbon atom is the less tightly bound.[3] So, we expect that this lone pair, which is not close to any atomic core in the molecule other than the one on which it is based (the carbon atom), will be very interested in any atom, molecule or ion which has regions of positive electrostatic attraction.

If we use the methods outlined in Chapter 6 to describe the expected shape of the boron hydride (BH_3) molecule, we find that the repulsions between the three bond pairs of electrons (there are no lone pairs) generate a symmetrical planar molecule with three equal H–B–H angles of 120 degrees. As we shall see later in Chapter 15, this molecule is highly reactive because there is very little electron distribution above or below the molecular plane. That is, there are regions above and below the plane containing the atomic cores in which *the net positive charges of those cores are not very well shielded by a negatively-charged electron distribution.* This molecule is looking for electrons in order to satisfy overall electrical neutrality in this region. Any molecule which has a suitable lone pair of electrons is capable of providing these electrons. In particular the CO (or the NH_3) molecule will do just that.

In exactly the same way that the lone pair of the ammonia molecule will attach itself to a proton — and for exactly the same reason — the lone pair of the carbon monoxide molecule will attach itself to the BH_3 molecule. The lone pair will fill the attractive region above the plane of the borane hydride. The resulting molecule then contains a new electron-pair bond, in which the two electrons which were originally a lone pair on the carbon atom of CO are now shared between the carbon atom and the boron atom to form a covalent B–C single bond.

---

[3]Remember that the oxygen atomic nucleus is less shielded than the carbon one, and so attracts — and holds — electrons the more strongly.

If we now think about the repulsions amongst the electron pairs surrounding the boron atom, there is a new situation:

- There are now *four* electron pairs surrounding this atom.
- All four pairs are bonding pairs and all repel each other.

It is easy to see, using the ideas of Chapter 6, that the combined molecule will have the three B–H bonds repelled out of their original planar configuration and the overall molecule BH_3CO will have the appearance of an 'inside-out' umbrella: the B–C–O being linear and the three B–H bonds being the umbrella ribs.

Here are contour diagrams of the CO lone-pair MO and the resulting B–C bond MO of BH_3CO.

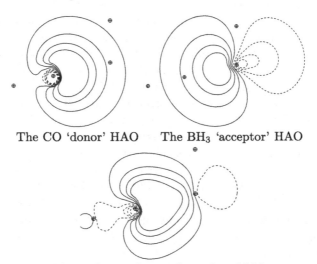

The CO 'donor' HAO   The BH_3 'acceptor' HAO

The carbon to boron dative bond MO

There are some particular points to be made about these diagrams:

- The 'contour slice' form of the contours has been used in place of the single 3D perspective, to emphasise the difference between the two HAOs. The slice is through the oxygen, carbon, boron and one of the hydrogen atoms.
- The lone pair on the carbon atom of the CO 'donor' molecule is much more compact and '*s*-orbital-like' than the 'acceptor' HAO on the boron atom of BH_3.

- To emphasise a point made throughout, one uses a boron atom HAO in the description of this bond because, in the bond, the shared electrons are close to the boron nucleus and so must take up a distribution similar to a boron HAO, whether or not the boron HAO is occupied in the BH_3 molecule.

10.1.3   *CO and transition-metal atoms*

The CO molecule has a specific capacity to be attracted to regions of positive electrostatic field surrounding transition metal atoms and ions. These metals have incompletely-filled $3d$ levels and this manifests itself as a region of space in which the atomic core is not completely screened by the electrons in $3d$ AOs. Carbon monoxide — as well as many other lone-pair-containing molecules — will supply electrons to this region, and in doing so it will form strong covalent bonds to the metal atoms.

One interesting and important example of this affinity of CO for transition metals is the straightforward reaction which occurs when CO is passed over hot metallic nickel. The reaction is simply:

$$Ni + 4CO \longrightarrow Ni(CO)_4,$$

where the four CO molecules are tetrahedrally disposed about the nickel atom.[4] The importance of the idea of the CO molecule donating electrons to the Ni–C *bond* rather than to the nickel *atom* is brought out by this example. If the nickel *atom* were the recipient of eight electrons it would be the sort of ion that might be made in the interior of stars and in nuclear accelerators, Ni^{8-}! Other molecules involving the dative bonding of CO to transition metals have formulae like $M(CO)_6$.

The addition of carbon monoxide to transition metal atoms and ions and the chemistry of the resulting molecules is of enormous chemical and industrial importance. For example, the vast majority of the ethanoic (acetic) acid made on the planet is by means of the simple-looking reaction:

$$CH_3OH + CO \longrightarrow CH_3CO_2H.$$

However, this reaction cannot be persuaded to occur except in the presence of rhodium or iridium salts and, when it does occur, produces ethanoic acid of greater than 99.9% purity. The elucidation of the mechanism of

---

[4]This is not so obvious as it is for the tetrahedral shape of the methane molecule because of the larger numbers of electrons involved.

the reaction shows that it involves the migration of a CO group originally attached to the metal atom!

10.2 The Dative Bond: Summary

In our considerations of the processes which occur on molecule formation in Section 1.3 on page 6, we had to consider what we think *would* happen if, for example, we want to describe the bonding in the methane molecule, CH_4. But methane is not actually generated in the laboratory by bringing four hydrogen atoms up to a carbon atom. In fact, making methane from (solid) carbon and (gaseous) hydrogen is not at all simple. The processes we have considered are mental experiments, rather than actual chemical processes. The main concern has been to see how covalent bonds in such molecules can be explained qualitatively on the basis of these sorts of arguments.

In the case of dative bonds, as we have seen, it is often the case that the real chemical reaction involved in forming the bond is precisely the same as our mental picture of it:

- Ammonia molecules in the gas phase will actually attach themselves to the proton of gaseous hydrogen chloride molecules and remove it to form NH_4^+.
- So will water molecules in the liquid phase, to form H_3O^+.
- Carbon monoxide gas will 'stick to' nickel atoms to form $Ni(CO)_4$.

This sort of dative bond formation provides the simplest possible form of bond formation. Very often reactions like:

$$A + B \longrightarrow A \rightarrow B,$$

which occur with no intermediate steps, no catalyst, no solvent etc. indicate the formation of a dative bond. This is in marked contrast to molecules containing covalent bonds. Indeed, the whole expertise of the synthetic chemist is to try to find ways of placing a covalent bond between two atoms or molecules: think of the special conditions and catalysts necessary 'simply' to add hydrogen atoms to a C=C double bond.

In this context it is worthwhile mentioning the method often used in organic chemistry to describe datively-formed covalent bonds. The general method used in organic chemistry to rationalise the movements of electrons — electron pairs always — during chemical reactions is by the use of curly or curved arrows. When used to indicate intramolecular electron movements, the arrows have to be curved to avoid interfering with the structural diagrams of the molecules involved. But there is, of course, the possiblity of chemical reactions involving the donation of electron pairs between molecules. In these cases, the same method is used: a curly arrow indicating the donation of a pair of electrons by one molecule (or atom) to another. Thus, the difference between the notation used here[5] simply boils down to whether the arrow indicating this process is straight or curved. In this case there is no danger of confusing the lines in the structural diagrams but, presumably, the arrows are also curved for reasons of consistency. Here are the two descriptions of the formation of the dative bond between CO and BH_3:

So, both the *formation* of dative bonds and the explanation of the mechanism of their formation are much more direct than for covalent bonds. However, in order to explain why some molecules have directed lone pairs of electrons at all, we have had to invoke their covalent bond structure and the mutual repulsions between electron pairs. It is these intramolecular electron repulsions which give the lone-pair electron donors their characteristic directional properties.

10.3 Assignment for Chapter 10

Dative bonds are the source of total disagreement in two large branches of chemistry. Organic chemists simply deny their existence on the grounds that, in the molecules they are mostly concerned with — consisting mainly of the elements H, C, N and O — the bonds, *once formed*, are not always

---

[5] And by inorganic chemists.

distinguishable from 'ordinary' chemical bonds. In fact, in the text quoted in Section 11.5[6] we find, on page 116,:

> Forget 'dative bonds' and stick to σ and π bonds.

In contrast, in almost all of transition-metal chemistry and much else in inorganic chemistry, the dative bond is a main type of electronic substructure invoked.

This is an intolerable situation. Try to form an opinion about this matter and how it might be resolved to avoid a permanent schism in chemical thinking. Discuss this problem with your organic and inorganic tutors[7] and see if there is any common ground. A useful starting point might be in discussing the mechanisms of the organic reactions which are catalysed by transition-metal compounds; often initial dative bonding generates products with new σ and π bonds.

---

[6] *Organic Chemistry*, by J. Clayden, N. Greeves, S. Warren and P. Wothers (OUP 2001) p. 549. This book is chosen since it is widely thought to be one of the best organic texts available, not from any disrespect to its able authors!

[7] Not at the same time, of course!

Chapter 11

Delocalised Electronic Substructures: Aromaticity

So far all the environment-insensitive electronic substructures we have met have been groups of electron pairs of various types: inner shells, bond pairs and lone pairs. A closer look at some organic molecules reveals that we must broaden our outlook and admit substructures with more than two electrons to our collection.

Contents

11.1 The Benzene Molecule

Benzene is a compound of enormous importance: commercially, chemically and in its contribution to the understanding of the chemical bond. The story of Kekulé's realisation — in a dream — that the molecule C_6H_6 is *cyclic* is part of the folklore of chemistry. However, the realisation that the molecule took the form of a regular (planar) hexagon of carbon atoms, each of which has a hydrogen attached, immediately poses new chemical problems:

- Why is the molecule planar?
- Why is the molecule a *regular* hexagon — with equal sides — when three of the C–C bonds are single and three are double? In other organic molecules a C=C bond is significantly shorter than a C–C bond. What is more, the C–C bond length in benzene lies *between* those of C–C and C=C bonds.
- Why is there only *one* molecule which has the hydrogens of two *adjacent* carbon atoms replaced by a substituent? If the above structure were correct, there would be two of these:

one in which the substituents are separated by a C–C bond, and the other where there is a C=C bond in between.
- Why is the total binding energy — the difference in energy between the benzene molecule and six carbon plus six hydrogen atoms — so large? This binding energy is much greater than that expected from three C–C bonds, three C=C bonds and six C–H bonds.

These facts, and many other similar ones, led to the realisation that, in the benzene molecule, *all the bonds between adjacent carbon atoms are equivalent.* In a world where the chemical bond had become identified with *localised pairs* of electrons this was a major challenge to theoretical chemistry.

There were other puzzles in the chemistry of benzene. In particular it was found that a benzene molecule in which one of the H atoms was replaced by another atom or functional group, the chemical nature of that substituent affected the *position* where another substituent would replace a second hydrogen atom! So, for example, in the molecule:

- If X is the hydroxyl group OH (phenol) then reacting phenol with bromine dissolved in cold carbon disulphide gives:

in about 85% yield.

- If, however, X is the nitro group NO_2, then reacting nitrobenzene with bromine requires much more violent conditions — about 140° C — and gives about 75% of

where, for the moment, the original Kekulé structure is used in spite of its now obvious deficiencies.

These chemical properties of benzene present a serious challenge to the theory of molecular electronic structure we have developed. It is clear that something new is happening in benzene which cannot be accounted for by the environment-insensitive substructures considered so far.

This is not an isolated phenomenon confined to benzene. It is found in a wide class of compounds which all share a common property:

Molecules which contain a continuous set of atoms bonded in a linear fashion, or in a ring in which the atoms of this

set are joined by *alternating* single and double bonds, show
properties which are not explicable by using the properties
of singly- and doubly-bonded atoms. They have, to a greater
or lesser extent, some of the properties which are epitomised
by the benzene molecule.

Some typical examples are: naphthalene, azulene, butadiene, pyridine,
the allyl radical.[1]

Perhaps the easiest way to begin to understand the structures of these
molecules is to look briefly at the earliest attempt to explain their properties using the 'conventional' substructures which we have met already.
The second and third of the items in the list at the start of Section 11.1
were the most puzzling: the fact that benzene is a *regular* hexagon and the
existence of only *one* 1,2 disubstituted benzene.

The explanation which appealed to chemists was that the single and
double bonds in benzene were swapping places very rapidly, so rapidly that
it was impossible to measure the lengths of the individual single and double
bonds. One simply gets an *average* value for both. The same explanation
would also work for the second puzzle. If the single and double bonds were
swapping places very rapidly, then separation of the two isomers would be
impossible. So the explanation was that the two possible structures existed,
but the expected properties based on only one of the structures are not
seen because of the rapidity of the swapping between two structures
like:

$$
\begin{array}{ccc}
& \text{H} & \\
\text{H} & & \text{A} \\
& & \\
\text{H} & & \text{B} \\
& \text{H} &
\end{array}
\qquad
\begin{array}{ccc}
& \text{H} & \\
\text{H} & & \text{A} \\
& & \\
\text{H} & & \text{B} \\
& \text{H} &
\end{array}
$$

This explanation is certainly acceptable:

(1) If one ignores the last of the properties listed above: the mutual binding
    energy of the six carbon atoms is much greater than three C–C bonds
    and three C=C bonds;

(2) If it is not known that atoms and molecules are composed of charged
    particles (electrons and nuclei).

---

[1] Look up the structures of these molecules in your organic chemistry text-book.

The objections are:

(1) This greater binding energy is simply baffling;
(2) If the single and double bonds are rapidly changing places this must mean that the electrons which comprise those bonds are moving about — oscillating — rapidly. Now, it is well known what happens when charged particles oscillate: they emit radiation.[2] If the electrons are oscillating they are emitting energy of radiation; where would this energy come from? It must come from the motion of the electrons and so the electrons would 'run down' over time. Further, the emitted energy would be detectable as light or heat or radio waves of some kind.

None of this happens,[3] so the explanation must be wrong!

11.2 Delocalised Electrons

If we look at the possible electronic structures for the benzene molecule it becomes immediately clear that it is the π structure where the problem lies. There is no problem at all with the σ system; it is analogous to the σ system of ethene discussed in Section 7.3:

- The C–H bonds are each described by localised MOs, comprising a linear combination of a polarised HAO on carbon and a $1s$ AO on hydrogen.
- Similarly, the σ C–C bonds have MOs composed of a polarised HAO on each carbon atom.

Both of these bonds are localised because the HAOs have strong directional properties along the relevant bonds.

However, if we take a portion of the benzene ring — say four carbon atoms — and straighten it out for clarity, we can see that there is a problem with the (potential) π-bonding. Looking at these four atoms 'from the side' (i.e. in the plane of the benzene ring) we see where, if we use the first contour diagram as an aid to clarify the overlapping, it is plain that *there*

---

[2]This is what is happening at the top of a TV transmitter or a mobile 'phone mast: electrons are being made to charge up and down a conductor and emit the signal we receive on our TV or 'phone.

[3]Check this for yourself. Put a bottle of benzene in a dark place and see if emits light, heat or interferes with your TV or 'phone and cools down.

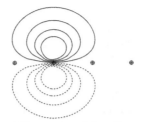

A 2p π-AO on one of the C atoms

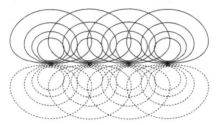

All four of the 2p π-AOs one on each atom

is no unique way of forming localised π *MOs* by taking linear combinations of the C-atom π (2p) AOs. This is the source of the difficulty:

> Obviously, looking at this fragment of the benzene ring, *we* can imagine forming two CC π MOs by forming linear combinations of the 2p AOs between atoms 1 & 2 and 3 & 4 *or* between atoms 2 & 3. If we put this in the context of the complete ring, we can imagine forming the two Kekulé structures by forming MOs with the two possible sets of partners. But, the electrons don't 'know' anything about this. They are just subject to Coulomb's law, quantum mechanics and the Pauli principle and, as such, will re-arrange themselves to minimise their energy *whatever we imagine they should do.*

Now is the time to use the clues from the experimental geometry of the benzene molecule. If the CC bond distances are all the same then:

(1) This does not change the *qualitative* nature of the σ bonding; it simply means that the C–C (σ) bonds are a little shorter than is usual for a typical C–C bond;

(2) There is absolutely no way to choose a sensible *localised* π-bonding scheme for the benzene molecule.

What has happened is that our cherished model of molecular electronic structure consisting of electron-*pair* substructures is not adequate for the benzene molecule and, of course, all those other molecules containing alternating single and double bonds.

However, whatever *we* think about it, it is clear that the electrons still have to obey the laws of physics: Coulomb's law, Schrödinger's mechanics and the Pauli principle. So we can, as usual, combine these with the main plank of the *chemical* model — use of (H)AOs — to calculate the electronic structure but, now, *without* using the idea of 'electron-pair' localised π bonds, since it is clear that this is inappropriate. Thus, we perform our usual method for the valence electronic structure:

(1) Use a set of polarised HAOs on each carbon and a $1s$ AO on each hydrogen (in the plane of the carbon ring), to form the C–C and C–H σ bond MOs.
(2) Use linear combinations of the π-type $2p$ AOs of the carbon atoms in the carbon ring to form the MOs of the π-electron substructure. Since we do not know the form of these MOs we simply let the calculation form the best description of them in terms of linear combination of the '$2p_\pi$' AOs.

The difference between the approaches summarised in 1 and 2 above is that we have not used a 'chemical model' for the π structure, except that, in line with our usual assumption of page 68, it can be expanded in terms of the separate-atom $2p$ AOs. In the first case, the expansion of the σ MOs is just:

$$\psi_\sigma = c_1\phi_1 + c_2\phi_2,$$

where ψ_σ is a localised σ-bond MO, c_1, c_2 are numerical coefficients and ϕ_1, ϕ_2 are HAOs. But, in the second case:

$$\psi_\pi = d_1 2p_{\pi1} + d_2 2p_{\pi2} + d_3 2p_{\pi3} + d_4 2p_{\pi4} + d_5 2p_{\pi5} + d_6 2p_{\pi6},$$

where, this time, $2p_{\pi1}, 2p_{\pi2}, \ldots, 2p_{\pi6}$ are the π-type $2p$ AOs on the six carbon atoms, $d_1, \ldots, d_6$ are numerical coefficients, and ψ_π is one of the three required MOs for the six electrons of the delocalised π substructure.

What we see when this is done is the expected σ-bond structure for the 'σ framework' of the molecule, plus a set of π MOs which are 'spread out' — delocalised — over all the atomic centres which have a π-type $2p$ AO. Here is a contour diagram of the lowest one of benzene, using the same four-atom fragment as in the diagrams immediately above:

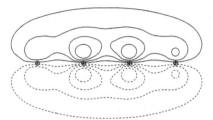

The lowest-energy delocalised π MO

The actual shapes of the contours for these MOs are rather hard to indicate for all the usual reasons, plus the fact that the molecule is rather large, and the delocalised MOs are spread over the whole carbon ring. Here is a view of the lowest-energy MO 'partly from above', looking down on the plane of the molecule at a slight angle:

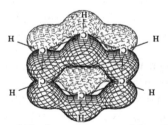

Perspective of the lowest-energy delocalised π MO

The electron distribution due to this MO is qualitatively similar in appearance, except that, being a *probability* distribution it is everywhere positive.

The other two π MOs — there are six electrons[4] in the π system of course — occupy MOs which, although spread out over the molecule, are

---

[4]C_6H_6 has $6 \times 6 + 6 \times 1 = 42$ electrons in total: six carbon atomic cores, six C–C σ bonds and six C–H bonds account for 36 of these.

not so symmetrical. These two MOs are actually *degenerate*,[5] although this is not obvious from their contours. Here they are in a similar perspective:

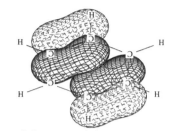

One of the pair of degenerate π MOs

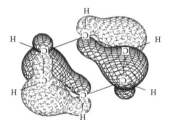

The other degenerate π MO

These orbitals and their associated electron distributions are descriptions of the new type of electronic substructure which occurs when there are a set of atoms in a molecule whose 'classical' valence structure would be written as contiguous atoms in a chain or a ring, joined together by alternating single and double bonds. Benzene is, in a real sense, the model for the ring compounds. The simplest 'straight-chain' molecule of this type is butadiene, with four π electrons and therefore two delocalised π MOs.

11.3 Environment-Insensitive π Substructures?

Considering the view taken in this work about the nature of the electron distribution in molecules — that they are formed of environment-insensitive

---

[5] Remember, in quantum theory 'degenerate' simply means 'having the same energy'.

substructures — we must see if these new delocalised substructures fit into our scheme.

Perhaps the first thing to say about the contour diagrams above for the degenerate pair of π MOs is that they look, at first sight, like *two* separate MOs: in one case like a pair of 'ordinary' π bonds, and in the other like a pair of π bonds extending over three atoms. This is misleading, because *both* of them are *single* MOs. In both cases, when singly occupied, they give (when squared) the overall probability distribution of that *single* electron which is indivisible; there cannot be 'half an electron' in each part of the MO. So *all three* of these π MOs are spread out over the whole carbon framework.

Secondly, one might expect that this six-electron delocalised π system would be similar across molecules containing a benzene ring: molecules with one or more of the hydrogen atoms replaced by functional groups. So, for example, we might expect similar π MOs in molecules with alkyl groups in place of the hydrogens:

etc., where, taking on board the discussion of the last section, a new notation has been used for the bonding in the benzene carbon framework: the localised σ system of single bonds is retained, but the central circle replaces the misleading alternating single and double bond diagram for the π structure. This expectation turns out to be satisfied, and, in general, provided the substituents:

- Do not themselves have a delocalised π structure adjacent to the benzene ring, and
- Are not very polar.

In such substituted benzenes, the structure of the benzene π system is similar from molecule to molecule.

However, there are two (related) properties of the π system which tend to break down its insensitivity to environment change:

(1) The π MOs are — as is usual for all π MOs delocalised or not — not as tightly bound as the σ MOs. The σ electrons are tightly held and rather difficult to move around the molecule. For example, substances composed of saturated molecules will not conduct electricity.

(2) The fact that the MOs are delocalised makes their electron distribution susceptible to being *polarised*. One can easily see that, if an electron is moved around the ring, the net attraction it experiences from the atomic cores and the σ electrons does not vary very much.

Both of these properties make benzene and similar compounds undergo reactions which are not found in the chemistry of molecules containing electron-pair single or multiple bonds and lone pairs. Their characteristic behaviour is due to the fact that the delocalised structure enables *transmission of influence* over comparatively long distances.

Consider a molecule in which the bonding is composed of electron-pair substructures, say a chain of carbon atoms with various hydrogen atoms or other functional groups attached. If one of the attached groups is strongly electron-attracting, this will polarise the other bonds which attach that carbon atom to other atoms or groups, making these bonds slightly (more) polar. This change in polarity will, in turn, polarise other bonds which are one atom further away from the original carbon. This process will continue through the whole connected chain. But the effect is opposed by the attraction of the atomic cores which hold the electrons in their original bonds and, typically, the presence of the electron-attracting group is only felt for one or two bonds. Now, consider what happens if one of the hydrogens in benzene is replaced by a strong electron attractor. The electrons in the *whole delocalised π system* are relatively free to move, and we might expect that the effect of the substitution will be felt over the whole extent of the delocalised π system with much less resistance than that in a molecule consisting of electron-pair substructures.

There are some celebrated cases which show this effect in a very marked fashion, not so much in their chemical properties, but in their physics:

Graphite is the ultimate in delocalised π electronic *ring* structures. It is composed of flat sheets of benzene rings fused together in two dimensions. The flat sheets are held together in the graphite crystal by relatively weak, basically intermolecular forces between the π systems outside either face of the planes. So, the whole everyday-sized crystal is a series of delocalised π structures. If we apply an electron-attracting 'group' — e.g. the positive pole of a battery — to one side of a graphite crystal, we can see the effect billions of CC bond distances away. In short, if we put a piece of graphite in a circuit with a source of DC voltage, a current flows; the electrons are free to move throughout each of the planar sheets of delocalised π structure.[6]

Polyacetylene (polyethyne) is, similarly, the ultimate in *linear* π delocalised electronic structures. It consists of long chains of C–H units joined together by alternating single and double bonds, so that the length of the delocalised π system is limited only by the ingenuity of synthetic chemists to make these chains. This substance also conducts electricity, but not nearly so well as graphite, precisely because is composed of many separate chains.

With these extreme cases in mind, we can now see that, if we take benzene and attach to it another similar structure, we will generate a new molecule which also has a delocalised π structure. But this structure will be different from *both* the constituents. For example, taking a benzene molecule and joining it to a butadiene chain in two possible ways will give the two molecules:

where only the carbon skeletons are shown and the traditional formulae are used to illustrate the alternation of bonds. Clearly, in each case, the product molecule will have a delocalised π electron structure spread over ten atoms, but they will be different from each other *and* different from the π systems of benzene and butadiene.

Thus, delocalised π systems are not 'portable' like the electron-pair structures which we have been discussing in earlier chapters. While each *individual* such system is relatively insensitive to changes in its environment — the π system of benzene or of naphthalene is not disrupted by substitution of the hydrogen atoms for functional groups — but it looks as if there are as many delocalised π substructures as there are ring systems or chain lengths. However, these delocalised π structures do share common *types* of property, principally the ability to carry 'information' over long molecular distances: directed substitutions in chemistry.

---

[6] Incidentally, the reason why graphite is so slippery — it is used as a lubricant — is because the flat sheets of benzene skeletons slide over each other so easily.

11.4   Nomenclature and Summary

It is clearly impossible to continue calling the molecules which we have been looking at in this chapter 'molecules with a ring or chain of alternating single and double bonds', a more compact name is needed. There are two such names which do not quite mean the same thing: one descriptive and one, which applies to a subset, historical.

First the historical name. This name derives from the fact that many of the common derivatives of benzene have distinctive smells and were called 'aromatic' for that very sensible reason. When it was realised that these molecules had distinctive properties which we have been discussing, the term aromatic was applied to certain types of ring structures which are much more tightly bound than expected on general chemical grounds, although, of course, not all such molecules have a pronounced — or any — odour. Although these ring structures do have particular electronic properties, the term 'aromatic' is now used rather loosely by practical chemists simply to mean 'more stable than we would think'.

The more descriptive term is 'molecules with *conjugated* double bonds', where 'conjugated' has the principal one of its many meanings: 'joined together'. If we take this 'joined together' to mean 'joined together by single bonds', then the term is satisfactory. In fact, as is usual, one usually simply talks[7] about 'conjugated molecules' or some convenient abbreviation.

However, while all aromatic molecules are conjugated, not all conjugated molecules are aromatic. Strictly, the term aromatic should only be applied to those molecules which have at least one delocalised π electronic structure over one or more closed rings of atoms, and this structure contains $(4n + 2)$ electrons for a positive whole number n. This includes both benzene and naphthalene and their derivatives, for example, but not butadiene, which is conjugated but is neither a ring system nor do its four π electrons fit the $(4n + 2)$ 'rule'. The derivation and detailed theory of aromaticity is outside the scope of what is attempted here, and a study of Hückel Molecular Orbital theory is a good entry point to understand these matters.

The question of why conjugated and, *a fortiori*, aromatic molecules are much more stable — more tightly bound — than would be expected simply by summing the energies of six C–H bonds, six C–C bonds and six C=C bonds, is not so simple to answer in terms of the ideas we have

---

[7]Among the chemical *aficionados*, of course.

been discussing. The full explanation requires some more quantum theory. If one solves the Schrödinger equation for electrons simply confined to a region of space (i.e. not necessarily in a molecule attracted by atomic cores), it turns out that, the larger the region of space in which the electron is free to move, the more stable it is, and it is this effect which is happening in conjugated molecules. The simple fact that electrons can 'spread out' from the localised π-bond structure makes the molecule more stable.

In summary, the delocalised π electronic structures provide an exception to the rule which has been the guiding principle for the earlier work, in the sense that although individual such structures are indeed stable and insensitive to changes in their environment, they are not 'portable' in the same very general sense that, for example, a C–H bond, a nitrogen lone pair or a C=C bond are. Indeed, one can join together molecules with no conjugated double bonds and produce molecules which do have spectacular new properties, as we have seen in the case of polyacetylene.

One final point: in discussing the likely properties of ordinary π bonds on page 139, we noted that, like lone pairs, the electron distribution tended to be away from the atomic cores and, because of this, would be more easily attracted to positive charges or regions of low electron density. This remark obviously also applies to delocalised π electronic structures. We should not be too surprised if we find that molecules like benzene and butadiene tend to 'stick sideways' to positive ions, in particular to multiply-charged transition-metal ions.

11.5   Assignment for Chapter 11

The whole question of aromaticity presents something of a difficulty for the classical methods of organic chemistry worked out in the 1920s by Robert Robinson, before Schrödinger's quantum theory was available. In particular, the organic chemist's view of reaction mechanisms is wedded to the idea that charge transfer in molecules occurs via the movement of electron *pairs* illustrated by curly arrows. Thus an explanation of the novel reactions due to conjugated delocalised systems is something of a challenge.

Here is a quote from a very popular organic text:[8]

'With some aromatic compounds, such as naphthalene, it [the depiction of the π electronic structure as delocalised or localised] does not mean that we think the double bonds are localised but *just that we need to draw curly arrows.*' [my emphasis]

Compare the scientific thinking here with Osiander who, in 1542, wrote what he clearly thought was a helpful introduction to Copernicus' revolutionary theory of the solar system with the sun (not the earth) at the centre:

'For my part I have always felt about hypotheses that they are not articles of faith but bases of computation, *so that even if they are false,* it does not matter, provided that they exactly represent phenomena.'

This approach is now very common, particularly in modern particle physics where the actual physical interpretation of a theory is less important than the fact that it gives the right results.

Discuss the merits and drawbacks of using familiar and tested methods in the face of new developments in scientific methods. Is it better to have something simple which works and gives the right result or something which explains why? Draw on your own experience and be completely honest — what works for you?

---

[8] *Organic Chemistry* by J. Clayden, N. Greeves, S. Warren and P. Wothers (OUP 2001), p. 549.

Appendix G

Some Historical Considerations

Contents

G.1 Introduction

The existence and properties of systems of delocalised electrons was, surprisingly perhaps, the first kind of system to be studied quantitatively by the methods of quantum chemistry. It is surprising because of the *size* of these molecules — benzene has 42 electrons and naphthalene has 68 — which looks, at first sight, like a formidable computational task. Yet approximate model calculations on the π-electron structure of these systems was pioneered by Erich Hückel in 1931, long before computers were available. Impressive as this work was and, perhaps because it was *so* unprecedented, it has left a legacy which has its unfortunate aspects. To see this more clearly, it is necessary to outline the details of what is now universally known as the 'Hückel model' and why it was so successful.

G.2 The Hückel Model

- The most basic assumption was that only the electrons in the delocalised π-system would be considered; these electrons were assumed to move in the 'effective' field due to the nuclei of the molecule and the electrons in the σ-framework.

- The MOs of the delocalised system are expanded as a linear combination of the $2p_\pi$ AOs of the atoms taking part in the delocalisation (e.g. the six $2p_\pi$ AOs of the carbon atoms in benzene). These two assumptions are common to most models of the π-structure.
- Because of the lack of computing facilities, only a very few of the energy terms required in a full MO calculation were used — the ones which were kept in the calculation were either guessed or simply carried through the calculation. In particular, the repulsions between the electrons were simply ignored and the overlaps of the $2p_\pi$ AOs were neglected. All these approximations affected the qualitative and quantitative value of the results of such calculations.

The obvious question to ask, in view of the severity of the scheme outlined above is 'why does it give useful results at all?' In fact, the procedure was most successful for *hydrocarbon* aromatics since the limited number of AOs used often means that the MOs are not very polar, and there is only a small number of different energy quantities to estimate. If we recall that the electron density of a molecule must reflect the overall symmetry of the nuclear arrangement and, since these electronic structures are *delocalised*,[1] there is every chance that the individual MOs are also delocalised and therefore each MO individually will have the symmetry of the nuclear framework. This proves to be the case; the best description of the electronic structure of conjugated π systems is one in which (1) the MOs are delocalised — spread out over the whole π system and (2) each MO has the symmetry of the nuclear framework of the molecule; each one gives the same contribution to the electron density from AOs which are equivalent by molecular symmetry. In the model for conjugated systems — benzene — the *quantitative* forms of the MOs are completely determined by molecular symmetry; a quantitative calculation based on *any* approximation to the values of the energy parameters will give the right answer.[2]

This was at once both the strength and weakness of the model; the aromatic hydrocarbons have π-structures which are very uniform precisely because all the atoms in the conjugated chain are the same and, in ring systems particularly, their environments are either identical or very similar.

---

[1] See the discussion in Section 8.1.

[2] Provided the numbers reflect the symmetry of the molecule, of course!

Thus, the qualitative features of the MOs and their *relative* energies are largely independent of the approximations made in the Hückel model. When heteroatoms are introduced into the conjugated system the situation is quite different; it is necessary to guess the way in which the energy parameters differ from the carbon standard and, ultimately, can only be determined by fitting the results to experimental data which destroys the capability of the method to explain and predict.

In the early days of quantum chemistry these successes of the Hückel method were so impressive — remember that *calculations* of (parts of) the electronic structure of large, chemically interesting had never been done before!

It is, of course, not too surprising that the neglect of the large repulsions between the electrons would have an effect on the quantitative actual energies of the electrons. But, since these orbital energies were expressed in terms of atomic-orbital energies which were unknown, only the *relative* energies were relevant, i.e. only the energy *order* of the MOs was available. But this neglect of electron repulsion had a major *qualitative* effect on the way in which approximate MO theories were interpreted.

G.3 Commentary

The successes of the Hückel model led to the ideas and concepts being taken over into the description of the electronic structure of molecules, for which the approximations used legitimately by Hückel were not appropriate either quantitatively or, most important, *qualitatively*. In fact, the qualitative properties of the basic system of all quantum chemistry — the hydrogen atom — and the successes of the Hückel model were combined in the following way:

The hydrogen atom is an exactly-soluble problem in quantum theory; all its allowed energy levels and orbitals can be obtained by the solution of a single (differential) equation. The energy levels and their associated orbitals describe the ground ($1s$) level and all the other levels. An electron can 'reside' in any one of the levels. There is no question of electron repulsion being involved in the energy of a hydrogen atom since it only has one electron — its energy is completely determined by the attraction of the nucleus and its kinetic energy.

The Hückel model generates a set of molecular orbitals which are then occupied by electrons according to the familiar scheme. Since there is no electron repulsion included, the energies of the MOs are not affected as electrons are placed in these orbitals. The orbital energies are *fixed* in the same way as those of the hydrogen atom. But they are only fixed in energy because a large component of the contribution to their energy (electron repulsion) has been omitted.

In other words:

> In the Hückel model, the MO energies are independent of their occupation. It makes no difference to the energy of an MO whether it is occupied by two, one or zero electrons. Nor does it affect the energies of the other electrons in the molecule. The Hückel MO energies, once calculated, are indifferent to changes in the electron distribution of the molecule.

This has been generalised to become what one might almost call the 'Quantum Chemistry Standard Model' of qualitative molecular electronic structure:

- Molecular orbitals are spread over the whole molecule; in general, they have contributions from all the (valence shell) (H)AOs in the molecule. If the molecule has any elements of symmetry, the form of each MO must reflect that symmetry.
- The MO energies are all available for occupancy by electrons and their energies are not affected by occupation.

G.4 Consequences

The authority and simplicity of the Hückel model has led to a whole series of applications and extensions which are well outside the range of the approximations made by Hückel in his pioneering work. We must remind ourselves that it was a very approximate method, using a semi-quantitative approach, to the electronic structure of a *particular case* of the delocalised π-structure of aromatic molecules. What this model does *not* imply is:

(1) *All* MOs and electronic structures are delocalised — as we pointed out earlier, some are and some are not; the existence of delocalised MOs does not automatically imply that the electronic structure of the molecule that they are being used to describe is delocalised.

(2) In particular, the highest occupied MO does not necessarily describe a particular, recognisable individual electronic structure in the molecule;

it may do, if that electronic structure is delocalised, or it probably will not if the system involves σ structures (localised pair bonds or lone pairs). Each case has to be looked at individually, there is no easy rule on this matter.

(3) Similarly, but perhaps more emphatically, the lowest unoccupied MO does not necessarily mean that this MO will give the spatial distribution of electron(s) added to that molecule; again, it may do but it probably will not. Since the energies of these unoccupied orbitals are usually positive, there is always a lower-energy state, in which the molecule is ionised, plus a free electron (a free electron has energy near zero).

(4) The MOs cannot be regarded as being even approximately independent of being occupied, in either energy or general form.

Now that the accurate calculation of molecular electronic structures is routine, these simple ideas are being replaced by a more realistic qualitative picture. But the fact that the properties of the energies and AOs of the hydrogen atom which do not involve electron repulsion still provide a tempting model for all electronic structures, and the simplicity and convenience of the Hückel model reinforces this temptation.

Chapter 12

Organic and Inorganic Chemistry

The concepts and methods developed so far provide most of the tools required to understand the environment-insensitive substructures to be found in the vast majority of organic molecules. The question is: 'can these methods be relied upon to explain the complexities found in the world of inorganic chemistry?' Organic and inorganic chemistry both deal with complicated molecules, but these molecules are complicated in very different ways: organic molecules are complicated because of the vast number of combinations and permutations of a relatively simple set of electronic substructures; inorganic molecules tend to contain more complicated individual electronic structures.

Contents

12.1 Commentary on Results

We have now spent considerable effort in developing a theory of the simple types of electronic substructure (atomic cores, covalent bonds, dative bonds and delocalised π systems) and with these tools it is possible to give a qualitative account of the valence electronic structures of most *organic* molecules. Organic molecules are largely composed of bonds among carbon atoms, bonds between carbon and atoms from H through F, and some lone pairs and conjugated systems. The complexity of organic chemistry arises in the main, not from the complexity of these individual electronic substructures but from the fact that these simple substructures can be combined in a bewildering variety of ways.

However, a few minutes' consideration reveals that, once we step outside the world of organic chemistry, things are very different. We quickly find out that our carefully thought-out methods do not even provide a description of some of the most familiar laboratory compounds: what is the electronic structure of the molecules in those bottles labelled 'Nitric Acid' or 'Sulphuric Acid', for example? In fact, almost the whole of inorganic chemistry is a new challenge in valence theory. The only parts of inorganic chemistry which we can feel at all at home with are just the 'organic' pieces of inorganic molecules; for example, the organic parts of organo-metallic compounds. In particular, how can we cope with:

- The familiar oxy-acids HNO_3 and their salts (nitrates and, for example, carbonates, also very common compounds);
- Compounds of the heavier typical elements, where variable valency is the rule rather than the exception;
- The compounds of the rare gases;
- The huge variety of molecules containing transition metals,

which seem so much more 'wild' than the familiar organic compounds? In fact, depending on one's point of view, inorganic chemistry is either an Aladdin's Cave or a Pandora's Box of new electronic structures. Each of the subjects opens up a whole new area of molecular electronic structure where some of our valuable rules break down and involve new types of electronic substructures. These individual structures will be the subject of further chapters. For the moment, it is worth while looking at the electronic structure of a very familiar oxy-acid and some common anions. After all, there they are in glass bottles on all our laboratory benches. It just will not do if we cannot understand their electronic structures.

12.2 Nitric Acid and Related Molecules

First of all, what is the structure of the nitric acid molecule; which atoms are joined to which? By this is meant the structure of the individual molecule, not its structure in solution in water; we can come to that later. Rather surprisingly, HNO_3 is *flat*, all the atoms lie in the same plane:[1] the nitrogen atom is bonded to all three oxygen atoms and the hydrogen atom is bonded to one of these oxygen atoms. Of course, the action of the molecule as an acid involves the loss of this hydrogen atom to give the nitrate ion NO_3^-. If we recall the kind of bonding situations which are common for hydrogen, oxygen and nitrogen we have met so far, there is already a problem here:

- The likely structure of the $H-O-N$ connected fragment apparently presents no problem: the hydrogen and oxygen atoms are simply showing their typical valencies of one and two respectively and the oxygen will, presumably, be attached to the nitrogen atom by a single (σ) bond. However, experiment shows that, rather surprisingly, there is a barrier to rotation about this $N-O$ single bond which is not easy to explain, since the usual explanation of hindered rotation about a bond involves a double bond between the atoms in the bond.
- The real difficulty lies in explaining how the remaining two oxygen atoms are bonded to the 'central' nitrogen atom. After all, experience would suggest that each oxygen atom would form a double bond (one σ and one π) to an atom which is its only bonding partner (think of O_2, CH_2O discussed earlier). But this would lead to a nitrogen atom involved in *five* bonds, not the familiar 'three bonds plus a lone pair' met earlier.
- On the other hand, we could imagine that, instead of the usual appearance of an oxygen atom as 'two lone pairs and two bonds', it might choose to have three lone pairs and one bond, as it does, for example, in the OH^- ion. This would give the nitrogen atom its familiar rôle with three bonds and a lone pair, *but* at the expense of predicting that the molecule was not planar about the nitrogen atom.[2]

The fact that the molecule is actually *planar* presents both a problem and a possible way out:

- A simple extension of the idea of π bonding in which the structure $O=N=O$ was found would suggest that this fragment would be linear.[3]

---

[1] Or rather, very close to the same plane: the $O-H$ bond is tilted by about two degrees to the plane of the NO_3 group.

[2] Remember the VSEPR 'rules' in Chapter 6.

[3] A σ bond from N to each O and π bonds formed as linear combinations of two *perpendicular* $2p$ AOs on N.

- But, a planar structure suggests, by analogy with the aromatic structures discussed in Chapter 11, that a multi-centre delocalised electronic
substructure might be involved.

In the absence of any real evidence about the nature of the electronic
substructures in the nitric acid molecule, we must fall back on the 'brute
force' method used in our investigation of symmetry in Chapter 8: simply perform an LCAO MO calculation and see what comes out. Since the
nitric acid has no significant symmetry, any calculation we perform should
not have the sort of acute problems we encountered with H_2O and benzene, in the sense that there should be no 'artificial' delocalisation in the
calculation. We shall have to try it and see.

Using the most obvious model of the electronic structure of the HNO_3
molecule:

- The familiar σ bonds for the O–H and O–N structures and;
- Simply allowing the out-of-plane π structures to be a linear combination
of all the available AOs of π type (the $2p_z$ AOs, say, if the molecule is
in the $x - y$ plane)

gives the following molecular orbitals:

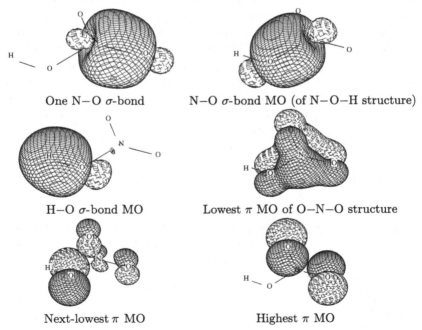

One N–O σ-bond   N–O σ-bond MO (of N–O–H structure)

H–O σ-bond MO   Lowest π MO of O–N–O structure

Next-lowest π MO   Highest π MO

The σ-bond MOs look very familiar and, as usual with localised electron-pair bonds, require no further comment. However, it is worth looking more carefully, at least qualitatively, at the forms of the π MOs and comparing these structures with the usual 'chemical' picture of the bonding in HNO_3.

(1) The lowest π MO has the same *general* appearance as the lowest π of the benzene molecule, in the sense that it is positive everywhere and clearly shows the 'spreading' of the π electron distribution from the atoms into the region between the atomic cores. This is obviously an MO which contributes to the bonding between the nitrogen atom and *all three* oxygen atoms. But it contains just *two* electrons and these two electrons are making a π contribution to three N−O, bonds unlike a conventional localised π bond of the type met earlier in (e.g.) ethene or methanal, where a pair of electrons are contributing to the bonding between just one pair of atoms.

(2) The highest π MO is just the opposite. It makes no contribution at all to the bonding among *any* of the four atoms of the 'NO_3 fragment', since it involves no contribution from the nitrogen-atom or one of the oxygen-atom AOs at all. Further, the contributions from the two oxygen-atom AOs are of opposite sign, suggesting that between these two atoms there is a *diminution* of electron density compared to the separate atoms.

(3) The intermediate π MO is of more complicated structure but again shows little sign of contributing to the π-electron density between the atomic cores, having a very small contribution from the AOs of the central nitrogen atom.

How can we give a qualitative summary of the contribution of the six electrons occupying these three π MOs to the bonding in the HNO_3 molecule?

The conventional 'organic chemistry' description of this molecule would be to say that it is a 'resonance hybrid' of the two equivalent structures like:

$$
\begin{array}{c}
O \\
\parallel \\
\underset{+}{N} \\
{}^{-}O \qquad O\!-\!H
\end{array}
$$

Can any sense be made of this idea using the above MOs?

First of all, the mystery of the restricted rotation about the N–O bond of the N−O−H fragment is solved: the delocalised π-electron substructure is making a *pi* contribution to this bond, and so it is not just a simple σ bond. It shows properties similar to the double bond (σ plus π) found in familiar organic molecules.

Second, it gives us a chance to think about what 'resonance' among structures which only differ in the distribution of double bonds actually means. If there are two (or more) possible electronic structures which differ only in the position of π bonds, then the 'true' structure of the molecule is said to be (or, better, is calculated to be) a linear combination of these structures, i.e. is a 'resonance hybrid' of these structures. What else can this mean *but* that the actual structure of the molecule contains a *delocalised π* substructure? The electrons are *not* distributed in localised π bonds, but are spread over the whole arrangement of atomic cores involved in the resonance structures. Of course, the prime example of this is the archetypical delocalised structure in benzene.

There is something to be said later about the various rules used in organic (and, to a lesser extent, inorganic) chemistry, but it is interesting to ask what the actual value of the idea of resonance is in describing the electronic structure of molecules. Perhaps its real function is to preserve the idea of the 'octet rule' in the face of evidence to the contrary.[4] It is also worth while to think about the VSEPR rules outlined in Chapter 6. The two atoms whose environments generate the shape of the HNO_3 molecule are the O atom of the H−O−N fragment and the central nitrogen atom. In fact, the principal fact which determines the shape of the molecule is not so much the repulsions amongst the localised σ electron pairs, but the delocalised π structure extending over the atomic cores of the NO_3 fragment. As noted above, it is this structure which holds the O−H bond in the plane of the rest of the molecule, as well as maintaining an overall planar structure for the NO_3 group. Once we admit *delocalised* structures (as opposed to simple, localised, π bonds), the simple rules based on the interactions of localised electron pairs need considerable modification.

12.2.1 *The nitrate ion NO_3^-*

Much of the chemistry of nitric acid is concerned with its dissociation in water into a hydrogen ion and the nitrate ion NO_3^-, both hydrated,

---

[4]The octet rule and its relationship to resonance will be discussed in Section 13.6.

of course. It is easy to see that, with the removal of the hydrogen ion
from nitric acid, what is left is the flat NO_3^- entity, in which the three
oxygen atoms are then symmetrically disposed about the nitrogen atom.
Calculations verify what one might expect: the molecule is just three N−O
σ bonds and a π substructure delocalised over all the four atoms of the
molecule.

Here is one of the three equivalent σ-bond MOs and the three delocalised π MOs:

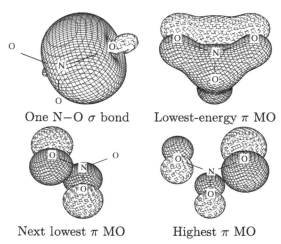

One N−O σ bond   Lowest-energy π MO

Next lowest π MO   Highest π MO

It is clear that these MOs are scarcely distinguishable from the ones of
nitric acid. The only small difference is the expected one that the N−O of
the N−O−H of nitric acid is slightly different from the other two, rather
than showing the complete symmetry of the ion.

What are we to make of these three delocalised MOs of these species?
The lowest-energy π MO is obviously contributing to the binding of the
molecule, but the 'bonding power' of the two electrons it contains has to
help bind all *three* bonds. The other two present more interesting cases: the
highest-energy one, having no contribution from the central nitrogen atom,
makes no contribution to the bonding, and the intermediate one does not
look as if it makes much contribution either. Indeed, in the case of the ion
there is no contribution from the nitrogen atom either, but there is a small
nitrogen contribution in the HNO_3 case.

What do we usually call orbitals in the valence shell of a molecule
which contain two electrons not contributing directly to the forces holding
the molecule together? In straightforward LCAO MO theory they are often

simply called 'non-bonding' MOs and the electrons they contain similarly non-bonding electrons. But this does not convey much information. So far, we have had a special name for such MOs and the electrons which they contain *if* they are on one centre and are basically (H)AOs: they are *lone pairs*. It would seem both sensible and meaningful, therefore, to call these π MOs in the NO_3^- ion at least *delocalised lone pairs*. Bearing in mind that the oxygen atoms of the nitrate ion each have two σ localised lone pairs, we can characterise the electronic structure of the nitrate ion (NO_3^-) as:

Atomic Cores: Four atomic cores, each consisting of the nucleus and a doubly-occupied atomic $1s$ AO;

Sigma Bonds: Three equivalent doubly-occupied $N-O$ σ bond MOs;

Delocalised (π) Bonding: The lowest-energy π MO illustrated above;

Sigma Lone Pairs: Two familiar 'sp^3' σ lone pairs on each oxygen atom not contributing to the molecular binding;

Delocalised (π) Lone Pairs: Two doubly-occupied delocalised π MOs not contributing to the binding of the molecule,

which shows some considerable complexity compared to the structures met so far in molecules composed of atoms from hydrogen through fluorine.

12.3 Carbonic Acid and Carbonates

The carbonic acid molecule H_2CO_3 is isoelectronic with nitric acid[5] and has the same planar shape. It forms both bicarbonates (salts containing the HCO_3^- ion) and the more familiar carbonates (containing CO_3^{-2} ion), and all three of these species are essentially planar.

12.4 Sulphuric Acid and Sulphates

The familiar oily liquid, which will react explosively with water, is both dangerous and interesting. Its overall structure is just four oxygen atoms around a central sulphur atom, with hydrogen atoms attached to two of the oxygens. We are again struck by the fact that, if we apply the octet

---

[5] Isoelectronic means 'having the same number of electrons' and is usually only used about molecules of similar overall geometry. So, for example, one would not say that the benzene molecule and the molybdenum atom were isoelectronic, even though both contain 42 electrons!

rule to the oxygens and the duet rule to the hydrogen, we are left with the sulphur atom being surrounded by *12* electrons. Furthermore, unlike nitric and carbonic acids, the sulphuric acid molecule is not planar; it is more akin to a tetrahedral SO_4 entity with the two hydrogens added to two of the oxygens, so destroying the exact tetrahedral symmetry. The sulphate anion, SO_4^{-2}, is actually tetrahedral so it cannot have the simple delocalised π electronic substructure found in the nitrate and carbonate anions. The sulphate anion will be investigated in more detail in Section 13.5.1.

12.5 Assignment for Chapter 12

This chapter has barely scratched the surface of the subject of the electronic structure of inorganic molecules. There is a vast and bewildering array of molecules of different types to be studied. Fortunately, inorganic chemists have, of necessity, become very good at collecting data about the structures and bonding in these molecules.

Use your inorganic text-book to have a look at the interactions between planar aromatic molecules, boranes and the transition metals, can you understand the qualitative descriptions of the bonding there? Do the ideas developed so far help you to understand these descriptions? Use your inorganic text to find a description of the bonding in some of these exotic species and see if you can find some common ground.

Chapter 13

Further Down the Periodic Table

All the discussion so far has concentrated on trying to explain the chemical bonds in molecules formed from the atoms H through F, the very 'top' of the periodic table. It should be the case that the same or similar arguments will apply to molecules containing atoms from 'lower down' the table, since we have concentrated on explanations involving the physical laws underlying the processes occurring on bonding, rather than rules involving the manipulation of orbitals. However, there are important differences between the electronic structures of atoms in the progression down any column of the periodic table. These differences are principally due to the fact that as the atomic number increases going down a column, the valence electrons are less tightly bound, and this has some important effects on their valence properties, since the valence of an atom is determined by how easy it is to polarise and share its electrons.

Contents

13.1   The Effect of Increasing Atomic Number

Any atom of atomic number Z is electrically neutral: it consists of a nucleus of charge $+Z$ and Z electrons each of charge -1.[1] But if we try to think of the difference between, for example, the electronic structure of the hydrogen atom $(Z = 1)$ and an atom of caesium $(Z = 55)$, we can immediately see some *qualitative* similarities and differences in the things which affect the distribution of the outer electron:

- In both of these atoms the net electric charge on the atomic core is $+1$.
- In both cases the outer electron occupies an s-type AO.

But:

- In the case of the hydrogen atom the *net* charge of $+1$ is just the *total* charge on the nucleus.

   In the caesium atom the *net* charge is the combined effect of the charge on the nucleus (55) and the 54 *electrons* in that atomic core.
- In hydrogen, the s-type AO is the simplest AO of all the $1s$ AO.

   In the caesium atom the s-type AO is the $6s$ AO.

If, for the moment, we ignore the fact that we know perfectly well that the $6s$ AO is of much higher energy than the $1s$, what effect would we expect these similarities and differences to have on the bonds formed by the two atoms? The most obvious difference must be the 'size' of the $6s$ AO compared to the size of the $1s$. The *repulsion* which the outer electron experiences from the other 54 electrons 'inside' the atomic core will force it to have its distribution mainly 'outside' that core where it experiences the combined *attractive* force due to the attraction of the nucleus and the repulsions of the core electrons. In short, the distribution — and therefore the AO — of the outer electron of caesium will be, on average, much further from the nucleus than that of the hydrogen atom. If this electron is further from the nucleus it must be *of higher energy* than that of the hydrogen atom: it is much easier to *change* the distribution of an electron in a $6s$

---

[1] In units of the charge on the proton, of course.

AO than one in a $1s$ AO. In particular is easy to remove the $6s$ electron completely from the atom than a $1s$ electron.[2]

Now, bond formation involves the polarisation and sharing of valence electrons. If valence electrons are easier to polarise and share the lower the atom is in the periodic table, we might expect significant differences in valence properties from the atoms of the first row of the table.

13.2 The Possible Demise of Lone Pairs

In Chapter 6 we considered the problem of why not all of the valence electrons in a given atom are shared with other atoms: why are not all the valence electrons involved in chemical bonds? For example, why is the typical formula of the hydrides of atoms of the first row of the periodic table not simply $XH_{(Z-2)}$, where Z is its atomic number? Sometimes this formula is correct (LiH, BeH_2, BH_3? and CH_4) but sometimes it is not (NH_3, OH_2, FH and 'NeH$_0$'). In the latter case one might expect NH_5, OH_6, FH_7 and even NeH_8!

The reason, as usual, is the particular balance in energy between:

- The lowering in energy obtained by sharing electrons between the attraction of two atomic cores.
- The raising in energy due to repulsions with other electrons and between atomic cores of atoms bonded to the same atom.

This balance is resolved by the existence of *lone pairs* of electrons when the number of valence electrons becomes greater than four for the atoms at the top of the periodic table. However, from what we have seen in the previous section, the balance between these factors might well be very different if the properties of the valence-shell electrons are significantly changed. If the valence electrons are less tightly bound and the distances between the atomic cores in a possible molecule are greater, it may well be that there is more energy to be gained in sharing more of the valence electrons than keeping them on the parent atom as lone pairs. Remember that molecules containing lone pairs are often very willing to share their lone pairs with other atomic cores if those atomic cores are attractive enough. This is the

---

[2]Notice the use of rather colloquial language here; electrons are *particles* and the electrons in the atomic core are just a swarm of such particles. We are using 'inside' and 'outside' (in quotes) as shorthand for 'in the space where the core electrons are, on average, most strongly distributed' etc.

source of dative bonding which we looked at in Chapter 10. If this balance is to be changed in practice it would require one or both of the following effects to be present:

(1) The energy gain obtained by sharing electrons between atom cores must be maximised; try to form bonds with atoms with strong electron-attracting powers;

(2) The energy loss generated by inter-electron and inter-atomic core repulsions must be minimised; use a central atom and, possibly, bonding partners with 'large' atomic cores and hence long bonds.

13.3   A Particular Case: Sulphur

The simplest molecule which the sulphur atom forms is its hydride, H_2S, which is the analogue of its periodic-table partner H_2O. But there are much more attractive bonding partners available than hydrogen, and the most attractive are the halogens which, for a given row of the periodic table, have the strongest electron-attracting power. The most strongly electronegative element is, of course, the 'smallest' halogen: fluorine. So, fluorine is the most likely bonding partner from the point of view of the criterion 1 above, but possibly not so attractive from the point of view of criterion 2. One might expect, for example, $S-F$ bonds to be shorter than $S-C\ell$ bonds. We might look around, then, for evidence for molecules like SX_n, where X is either F or $C\ell$, and n is greater than 2.

What is found experimentally is that there *are*, in fact, two such molecules: SF_4 and SF_6, where, as we might expect from Chapter 6, the number of shared valence electrons goes up in steps of two, as the two lone pairs of SF_2 are progressively shared, rather than the valence showing odd-number steps. The shape of the SF_6 molecule is very easy to predict by the rules of Section 6.3, since there are no lone pairs present. All six valence-shell electrons are shared in bonds. The molecule is therefore very likely to be the most symmetrical arrangement of six identical atoms around the central sulphur: the corners of a regular octahedron.

The shape of the SF_4 molecule is more of a challenge: four bond pairs and one lone pair. If the lone pair were not there, the preferred shape would be determined simply by the repulsions amongst four equivalent bond pairs; qualitatively the same as the methane molecule CH_4. We would expect a tetrahedral molecule. Since the lone pair *is* present in SF_4, the molecule will not be tetrahedral; it contains: four equivalent bond pairs and one lone pair. Remember the repulsion 'rules' from Section 6.3:

$$\text{LP with LP} \quad > \quad \text{LP with BP} \quad > \quad \text{BP with BP,}$$

where 'LP' means a lone pair and 'BP' means a bond pair of electrons.

So, there are six BP−BP repulsions, four LP−BP repulsions with no LP−LP repulsions, and the solution is not so obvious. To be sure we would have to know the actual numerical values of the two types of repulsion. What are the possibilities? With five electron pairs we can always make a start by making the approximation that they are all equivalent. In that case we would expect some form of 'bi-pyramidal' arrangement of the pairs; three pairs in a plane and the two remaining pairs above and below the centre of the plane, but it is not at all clear which pairs would be where when they are not all equivalent.

There are various refinements possible to the basic VSEPR rules, but such refinements tend to have the air of being used to fit the facts rather than being based on scientific grounds. Perhaps the best thing to do is simply to calculate the numerical size of the pair repulsions and see what comes out. It is useful to have a general idea of the approximate size of the electron-pair repulsion energies. For this, we can make the extremely coarse approximation that each pair can be modelled by a charge of two electron units and that these charges are about one Angstrom unit apart. In atomic units where the charge on an electron is -1 and one Angstrom is about two atomic length units, the repulsion energy between two such pairs is given by Coulomb's law:

Repulsion Energy = (Product of Charges)/(Inter-charge distance)

$$\approx \frac{(-2) \times (-2)}{2} = 2$$

where, of course, the energy is in atomic units of energy which, when scaled up by the Avogadro number, gives quite a large amount of energy per mole (two atomic units is about 5246 kJ per mole). However, just for purposes of comparison the units do not matter, provided everything is in the same (atomic) units.

When the optimum shape of the SF_4 molecule is calculated, the shape is found to be a distorted triangular bi-pyramid, in which the central 'base' of the bi-pyramid is formed by two (equivalent) S−F bonds and a lone pair with the two other (equivalent to each other but different from the first two) above and below the triangle. A schematic diagram is given below.

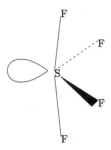

The mutual repulsions (in atomic units) of the five valence electron pairs (four S−F bonds and one lone pair) are summarised in the table below:

Repulsion energy between electron pairs in SF_4

	Bond 1	Bond 2	Bond 3	Bond 4	Lone Pair
Bond 1	—	1.34	0.93	1.34	1.50
Bond 2	1.34	—	1.34	1.29	1.43
Bond 3	0.93	1.34	—	1.34	1.50
Bond 4	1.34	1.29	1.34	—	1.43
Lone Pair	1.50	1.43	1.50	1.43	—

Bonds 2 and 4 are the ones in the same plane as the lone pair, and Bonds 1 and 3 are the ones above and below the plane. These numbers represent huge amounts of energy (remember one atomic unit is around $2600\,kJmol^{-1}$). But there are equally large other factors involved in the interactions between the nuclei and electrons involved. It is the net effect of all these forces, which largely cancel, that decides the shape of the molecule as we discussed in Section 6.3.

13.4   The General Case: 'Hypervalence'

With the example of sulphur in mind we can now generalise for the expected valencies of each of the columns of the periodic table below N, O, F and Ne, always bearing in mind that the higher valencies may well only be shown when the attached atoms are electronegative enough for the sharing of electrons to overcome the effects of electron repulsions. Perhaps the simplest way to summarise the *expectations* is in the form of a table:

Atoms	Lone Pairs	Valence	Likely Molecule
P, As, Sb	1	3	PH_3
	0	5	PF_5
S, Se, Te	2	2	SH_2
	1	4	SF_4
	0	6	SF_6
Cℓ, Br, I	3	1	$C\ell H$
	2	3	$C\ell F_3$
	1	5	$C\ell F_5$
	0	7	$C\ell F_7$
Ar, Kr, Xe	4	0	Ar
	3	2	ArF_2
	2	4	ArF_4
	1	6	ArF_6
	0	8	ArF_8

Some comments on the table are in order:

- H_2S and HCℓ have been written as SH_2 and CℓH simply for comparison purposes.
- The first member of each series of three atoms is used for the prediction 'likely molecule'. This may not be appropriate, since, as we have noted, the larger valencies are more likely to be shown for the heavier members of the groups.
- It has been assumed throughout that the valencies will all be 'used' in forming the relevant number of *single* bonds. That is, for example, the 'likely molecule' between S and F when S has a valence of six is shown as SF_6 rather than SO_3, where the S and O are joined by double bonds.

In looking to see if our predictions are correct the last two assumptions will have to be borne in mind.

We should include the most surprising prediction: that the 'inert' gases will not only form molecules, but will form a whole family of molecules containing several covalent bonds. Here is a selection of known molecules formed from the atoms near the bottom of the N, O, F and Ne columns of the periodic table:

Atom	Lone Pairs	Valence	Known Molecules
Sb	0	5	SbF_5
Te	1	6	TeF_4
	0	6	TeF_6
I	2	3	IF_3
	1	5	IF_5
	0	7	IF_7
Xe	3	2	XeF_2
	2	4	Xe_4
	1	6	XeF_6
	0	8	XeO_4

There are many others, some of 'mixed' type showing both double and single bonds. Notice that XeF_8 is *not* found experimentally but XeO_4 is known, in which, presumably, the xenon atom is showing a valence of eight if the oxygen atom is exhibiting its familiar valence of two.

There are similar results for the other columns of the periodic table. The higher valencies are all found but the very highest valencies tend to occur involving double bonds with correspondingly fewer bonding partners than a full set of single bonds: atoms involving eight single bonds are very rare, even with the most powerful electron sharer, fluorine.

13.4.1   *Single or double bonds?*

It has been repeatedly stressed that the formation of chemical bonds by an atom is always a balance between the energy gain obtained by sharing electrons with other atoms and the necessary consequence: the energy loss due to repulsion between the shared electrons and other electrons around that atom, other bonds and lone pairs on the atom. In the cases looked at in the previous section this balance is particularly precarious: as more and more electrons are shared by a given atom there is more and more 'crowding' around that atom. Each electron shared involves the formation of an electron-pair bond, that is, one additional electron in the vicinity of the given atom.

We have seen already that when the bonds are relatively short, as they are for atoms at the top of the periodic table, the balance is very much over to the side of 'fewer bonds and more lone pairs'. This is because, although there is energy to be gained by electron sharing, the repulsive forces between the shared electrons and between the bonded nuclei are too great to permit the formation of more than a few new electron-pair bonds. Going to the other end of a column of the periodic table we would expect that, sooner or later, even though the electron-pair bonds are longer and the repulsions between them are therefore weaker, there must come a point when so many new electron pairs are formed that the repulsion amongst them 'gets the upper hand' over the energy gained by forming those bonds.

One possible compromise here is simply to cut down the repulsions amongst the nuclei attached to a given atom by sharing more than one electron with that atom: in a few words, forming a double bond with one atom rather than a single bond with each of two atoms. This is clearly what has happened with the eight-fold valence of xenon: the repulsions among eight electron-pair bonds and among eight fluorine atomic cores are so great that even the strong electron-sharing power of the fluorine atom is not able to overcome them, and XeF_8 is not formed. But the much less 'electron-pair-crowded' molecule XeO_4 is stable.[3]

Talking about 'crowding' around an atom or molecule due to the presence of many electron pairs naturally raises the question of the 'steric effect'. It is worth a short discussion here, since it plays a rôle in the thinking of both organic and inorganic chemists.

13.4.2 The steric effect

The 'steric effect' is the name given to any phenomenon (usually changes of shape) arising due to the crowding together of chemical bonds, lone pairs or whole functional groups within a molecule. A familiar 'large-scale' example should make the nature of this effect very clear.

When we clap our hands together, what is it that prevents them from simply passing through each other? At the human scale of things the answer is obvious enough. It is the sheer 'solidity' and 'space-occupation' of our hands which makes them stop suddenly on collision; they are solid, not gaseous! But, at the submolecular scale things are very different. Remember electrons and nuclei are *tiny* particles; all the particles comprising both our hands would fit comfortably into the volume of a grain of sand if they were not electrically charged. The vast majority of what we perceive as the 'volume' of our hands is empty space, as empty as the space between

---

[3]Remember that each atom joined to the Xe atom also has unpaired electrons of its own to add to the crowding.

the planets in the solar system, to go back to a large-scale model for a moment.[4]

It is not the size and space-occupancy of the particles of which our hands are composed which makes our hands clap rather than simply gliding through each other, but the fact that these particles are *electrically charged* and therefore experience very strong interactions. The 'outer' regions of all atoms and molecules are composed of electrons in very rapid motion; so rapid that we can *imagine* them as effectively filling the space in which they move to form a distribution of charge. These electron 'distributions' repel each other very strongly via electrostatic effects and, to a lesser extent, through the action of the Pauli principle. It is these strong repulsions which we subjectively experience as the impact of solid objects.[5] Of course, it does no harm to continue to speak of the effect of crowding, or to call this phenomenon the steric effect, but we must remember that, like so many other named effects in chemistry, it is nothing more than short-hand for the combined effect of our three laws of nature:

- Coulomb's law (repulsion of electrons).
- Schrödinger's mechanics (continuous motion of electrons).
- The Pauli principle (additional repulsion of identical particles).

Like other named effects in the theory of molecules (e.g. the octet rule) and molecular interactions (e.g. van der Waals forces) it should not be thought of as independent of these basic laws; it is simply a convenient way of summarising their combined effects.

13.5   How to Describe These Bonds?

It is straightforward to see the physical reasons for the increases in valency as we go down the periodic table, but how are they to be described in terms of the *tools* which we (along with everyone else) are forced to use. That is:

How are we to describe the electron distributions in terms of the AOs of the constituent atoms of the molecule?

---

[4] If two solar systems were to 'collide' it would be a rare event indeed for two of the component planets to actually hit each other!

[5] You may need to convince yourself of this idea which seems an outrage to 'common sense'. If you do this, it will help with your understanding of the electronic structure of molecules.

The problem arises for exactly the same reason as the use of the $2p$ AO was necessary in LiH which was described in Section 3.2: the AOs which are *occupied* in the separate atoms may well not be adequate to describe the changed electron distributions in a molecular situation.

Remember the main reason for using AOs in the description of molecular electronic structure:

> The electron distribution close to an atom *must* be described to a very good approximation by atomic orbitals, because the electron distribution close to that atom is dominated by the strong attractive field of its atomic core.

There is no requirement in this general principle that the description should be confined to the AOs which are occupied in the ground state of an atom, as the lithium hydride example showed very clearly. Using only the occupied valence AO of the lithium atom ($2s$) would prevent us from generating a *polarised* distribution on the lithium atom and therefore an important physical process on bond formation.

The detailed situation is different, for example in SF_4 and SF_6; the problem here is that, if we were limited to using only the occupied valence AOs of the sulphur atom ($3s$ and $3p$), there simply is not the flexibility to describe electron distribution when the sulphur atom shares more than two of its six valence electrons in chemical bonds (four in SF_4 and all six in SF_6). If we recall the brief summary given in Chapter 2, the key facts here are:

(1) Since the energy levels available to electrons in atoms depend on $-1/n^2$, they get closer and closer together as n increases.
(2) For a given value of n, the s level lies lower than the p level, which in turn, lies lower than the d level: $E_{nd} > E_{np} > E_{ns}$.

The natural AOs to be used in trying to give an adequate description of valencies of four and six for sulphur is then, by analogy with the lithium hydride case, the AOs which are the next-higher in energy. These are, of course, the AOs which would be occupied in the excited states of that atom: the set of five $3d$ AOs. When this is done, calculations show very clearly the form of the MOs describing the electron distributions in (e.g.) the four S$-$F bonds in SF_4, and the single remaining lone pair and the six S$-$F bonds in SF_6 with no lone pair.

13.5.1 *A comparison: 16 valence electrons*

In the next chapter some of the *differences* between the properties of the typical elements and the transition elements will be highlighted. In this section the transition from the typical to the transition element can be noted. There are a number of molecules[6] with the general formula AO_4, a central atom (A) joined to four oxygen atoms often in a symmetrical tetrahedral arrangement. As we noted above, oxygen, while not such an electronegative (electron-attracting) atom as fluorine, does have the useful property of easily sharing *two* of its valence-shell electrons readily to form double bonds. This makes it a promising bonding partner which will enable another atom to show its full 'hypervalence'. It is strongly electronegative *and* there will only be half the number of atoms bonded to the target atom as there would be with the fluorine atom, so that the mutual repulsions between the bonded atoms are minimised. Three convenient examples are:

(1) A neutral molecule: osmium tetroxide OsO_4. Valence-electron configuration $6s^2 5d^6$.
(2) A singly-charged anion: permanganate MnO_4^-. Valence-electron configuration $4s^2 3d^5$ plus a single electron giving the negative charge.
(3) A very familiar doubly-charged anion: sulphate SO_4^{2-}. Valence electron configuration $3s^2 3p^4$ plus two electrons making up the charge on the ion.

The three central atoms — the A of AO_4 — are quite different; they differ by the row of the periodic table in which they occur and the nature of their valence-electron configuration. The corresponding three tetroxide entities have, however, one vital thing in common:

> They each have 16 valence electrons in total through the combination of the two easily-shared electrons of the four oxygen atoms, the valence shell of the central atom (S, Mn or Os), and being made up to 16 by the electrons providing the charge on the whole system: zero, one or two.

It is worth asking the question:

> Does the sharing of these 16 electrons generate a similar bonding pattern in the three AO_4 entities *in spite of* the

---

[6] Remember, 'molecules' includes ions.

very different nature of the central atom A, or is there a qualitative difference in the bonding?

Before presenting any analysis of the valence structures involved here, it is worth emphasising the usual models used. If we look at the chemist's picture of the bonding in these systems, what is very clear is that the models used to describe the electronic structure of the metal-containing entities is very different from the model used for the sulphate anion:

- The sulphate anion is usually said to have a configuration which is an equal linear combination — a 'resonance hybrid' of the type discussed in Section 5.5.1 — of structures which have two S−O single bonds and two S=O double bonds, with the negative charges being carried by the oxygen atoms involved in the S−O single bonds. Of course, the resulting overall structure must have tetrahedral symmetry, and this forces the combination co-efficients of the differently-bonded components to be equal.
- The electronic configuration of the Mn and Os species is described by a completely different picture. Transition-metal-containing molecules are typically described using MOs which are symmetry-adapted, and so each MO has the symmetry of the tetrahedral framework, as discussed in Chapter 8. Without going into the details of this analysis, this model gives a structure which contains eight electrons in bonding MOs, each of which is symmetry-adapted, and so does not say much about what is happening on bond formation. It does, however, suggest that there might be just *four* chemical bonds (two net bonding electrons per bond) — rather than the *six* in the sulphate anion — which are not localised.

It is perfectly straightforward to carry out standard MO calculations on these three systems and analyse the resulting overall electronic structure for substructures of the type used earlier. What emerges is an interesting confirmation of the ideas buried in the above, apparently unrelated, models.

SO_4^{2-}: The analysis gives the surprising result that this molecule has four equivalent double S=O bonds, which is something of a surprise since we only expect sulphur to have six bonds at most, as found in SF_6. However, when the bonds are looked at in detail, things begin to make sense. Here are the slice contours of one of the double bonds:

The localised MOs describing the double bond in SO_4

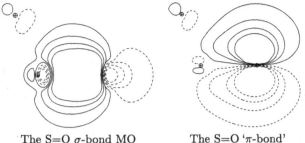

The S=O σ-bond MO   The S=O 'π-bond'

The 'π bond' is very asymmetrical, with only a very small electron density near the central sulphur atom. It is 'almost a π lone pair on the oxygen atom'. The actual total charges associated with the atoms are $+1.5$ on the sulphur atom and -0.87 on each oxygen, leading, of course, to a total charge on the ion of -2.0, as it should. The remaining valence electrons occupy a σ and π lone pair on each oxygen atom.

MnO_4^- and OsO_4: In both these cases, the analysis gives no hint of double bonds between the central metal atoms and the oxygens. The valence structure is much simpler: a single bond (Mn–O or Os–O) and, this time, three lone pairs on each oxygen atom: one σ and two π.

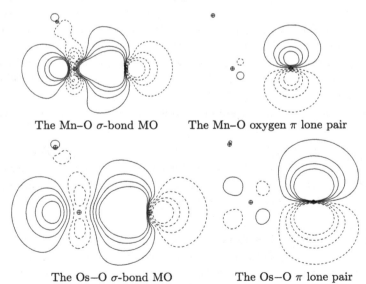

The Mn–O σ-bond MO   The Mn–O oxygen π lone pair

The Os–O σ-bond MO   The Os–O π lone pair

Notice that, in line with what was said in Section 7.7, the σ/π terminology is being used to refer to each bond separately; the π orbitals are perpendicular to the local bond and not to some plane of symmetry in the molecule. Both MnO_4^- and OsO_4 have the oxygen atoms joined by a single bond and three lone pairs: a structure which is also found in, for example, the OH^- ion, a very common species. What is just visible in the diagrams is that in the sequence:

$$SO_4^{2-} \longrightarrow MnO_4^- \longrightarrow OsO_4,$$

the S–O π system is partially attracted to the S atom; the same system in Mn–O looks entirely localised on the O atom, while in Os–O it seems to be slightly repelled from the Os atom.

While one can make a case for saying that the three σ MOs all look qualitatively similar, there is *one* striking difference which cannot be ignored:

> In the case of the two transition-metal to oxygen σ-bonds there is a large 'bulge' of electron distribution in a direction *away* from the Os–O and Mn–O bonds, which is not present in the S–O σ-bond.

This turns out to be important in many contexts, and will be discussed in more detail in Chapter 16. But, for the moment, this difference can be easily explained by the *detailed composition* of the MOs, in particular the difference between the sulphur atom HAO and the two metal HAOs, since the oxygen atom σ MO is much the same in all three cases (approx. *sp*). The slice contour diagrams give a clearer idea of the forms of the HAOs:

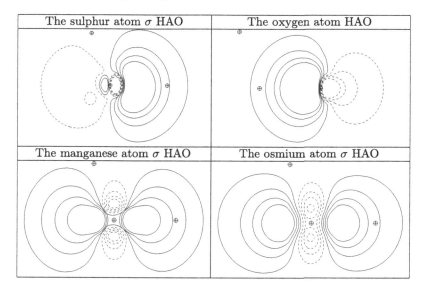

The sulphur atom σ HAO	The oxygen atom HAO
The manganese atom σ HAO	The osmium atom σ HAO

The reason for this striking difference is simply that:

> The sulphur σ HAO is composed of $3s$, $3p$ and $3d$ AOs in similar amounts, while the metal σ HAO is composed of $4s$ (or $6s$ for Os) and $3d$ ($5d$ for Os) with a very small amount (about 0.05%) of $4p$ ($6p$ for Os). The opposite 'lobes' of a p AO are of opposite sign, reinforcing the s AO on one side and partially cancelling it on the other. The opposite lobes of a d AO are the same sign, making a more symmetrical HAO with similar lobes on each side of the atom.

In the case of a tetrahedral molecule the lobe opposite to the O atom 'points' at no atom and has the appearance of 'half a lone pair'.[7] The situation is quite different, as we shall see, in the more common octahedral transition-metal molecules, where each lobe of the HAO points to an atom.

13.6   An Updated Summary

We are now in a position to give a fuller generalisation of the factors involved in chemical bond formation than the one at the conclusion of Chapter 6. The main points are identical to those of Section 6.5 — the 'Working Summary' — but can be made more general by including the conclusions reached in Section 10.2 about dative bonds, and the material of this chapter concerning 'hypervalence'.

To avoid too much page-turning, here are the conclusions of Section 6.5:

- A chemical bond is formed when the energy of a pair of electrons (one from each partner in the bond) can be lowered by being shared between the attractive force field of two atomic cores simultaneously, rather than each electron being attracted solely by its 'own' atomic core.
- When an atom is involved in several bonds with different atoms there is always repulsion between these bonds (between the electrons of the pairs and between the atomic cores of the bonded atoms).
- How many atoms a given atom can bond to depends on the balance between these two things:

(1) The energy lowering due to the sharing of electrons.
(2) The energy raising due to the repulsions among the bonds formed.

---

[7]Whatever *that* means.

The most obvious revisions are:

• To the first of these points, considerations of dative bonding mean that the proviso 'one [electron] from each partner in the bond' can be omitted, which would include dative bonds in the scheme.

• The third item needs slight expansion to make room for the discussion in this chapter to include the fact that, for a given atom, multiple bonding with bonding partners serves to reduce the 'steric' repulsion involved, and so tends to promote the maximum valence of that atom.

13.7 Assignment for Chapter 13

One of the interesting features of hypervalence is the fact that, in order even to have enough orbitals to describe these molecules and satisfy the Pauli principle, it is necessary to use AOs which are not occupied in the ground states of some atoms. We met this very early on in the discussion of the formation of a polarised HAO in the lithium hydride molecule and, to a lesser extent, even in the simple hydrogen molecule. In the case of lithium there are 'enough' AOs to give a formal description of the bonding — one AO per atom — but this description is inadequate. These examples highlight a point which has been emphasised throughout:

> (H)AOs are used in describing the electronic structure of molecules, not because they are actually *occupied* in the atoms of which the molecule is composed,[8] but because they are *suitable* for a description of these electronic structures.

You may be able to find in the older literature a different approach to explain the use of unoccupied AOs in describing the electronic structure of molecules. In fact, this method was used to explain why carbon has a valence of four when it only contains two unpaired electrons (the ones in the $2p$ AOs). The argument has a rather 'thermodynamic' feel; electrons are promoted from the ground-state-occupied into the unoccupied AOs during molecule formation. Obviously, promoting electrons requires energy to be supplied from somewhere, but this energy (and more) is regained when the bonds are formed.

What do you think of this method; does it represent a real process? How does this scheme tie in with the use of hybrid atomic orbitals?

---

[8] Although, of course, many are.

Chapter 14

Reconsidering Empirical Rules

The last two chapters have shown that the first of the simple rules developed to describe the valence of atoms — the octet rule — has only a limited application and cannot, therefore, be considered as a guide to likely molecular structures. Of course, the octet rule was proposed long before there was any clear understanding of the laws which govern the distribution of electrons. This chapter addresses some of the problems with this type of empirical rule revealed by the phenomenon of hypervalence and variable valence.

Contents

14.1 Limitations of the Octet Rule

It is becoming increasingly obvious that the octet rule, mentioned briefly in Chapter 2, is not very relevant to the electronic structure of inorganic molecules, particularly those molecules containing atoms with an atomic number greater than ten. It is perhaps time to look at the rule with a critical eye, see what evidence it is based on and to what atoms it *does* apply:

- First of all, the rule was introduced in an attempt to find common properties in the chemistry of the elements of the first two periods[1] of the periodic table: hydrogen through neon. But, of these elements, when it does actually apply it is only to carbon, nitrogen, oxygen and fluorine, with carbon being the most 'law-abiding' element of all. It is very uncommon to find a carbon atom in an environment where the octet rule cannot be made to work. There are some *classes* of exceptions to the rule for carbon as we have seen, the most important being those molecules containing delocalised π substructures (molecules like benzene and the carbonate anion). Fluorine, with its predominantly single bonding character, is also well behaved. But nitrogen and oxygen are rather unruly at times, as we have seen in the case of nitric acid and the nitrate anion.
- In the third and higher periods, where variable valency is the rule rather than the exception, the octet rule does not have much validity. Lower down the periodic table 'hypervalency' is common. The term hypervalency suggests that these are unusual cases but, in fact, this term is only used because the octet rule is not obeyed, *not* that it is at all uncommon. As we have seen, the octet rule is only 'obeyed' for one or two atoms, and is not even the commonest case for many other atoms.
- The transition elements are particularly unruly in this respect, and they have a separate 'rule' of their own — the 18-electron rule — to which there are many exceptions.

In summary, it does not seem sensible to use the chemical properties of a *very* small number of light atoms as a standard against which to judge the chemical behaviour of the other 88 elements of the periodic table. Perhaps 'hypervalency' should simply be called 'valency' if we want a term which applies to most atoms.

14.2   The Basis of the Octet Rule

To understand fully what the rôle of the octet rule is and how it has come to be regarded as the standard rather than just a property of a few

---

[1]The octet rule cannot be expected to be obeyed by hydrogen and helium; perhaps there should be a 'duet' rule for these two atoms.

light atoms, is is necessary to stress that the rule was produced some time before there was any theory of the energies and distributions of electrons in atoms and molecules. Remember the laws of nature which *do* govern the distributions and energies of sub-atomic particles, introduced at the very start in Chapter 1:

(1) The law of force operating between charged particles (Coloumb's law);
(2) Schrödinger's mechanics;
(3) The Pauli principle.

Note that the octet rule does not mention them or, indeed, *any* law of nature. But, since the rule *has* proved useful in looking for system in the chemistry of the elements, it must in some way or another be *derivable* from the laws governing the distribution of electrons.[2]

The relationship is fairly obvious:

> When Schrödinger's mechanics is applied to the case of atoms (Coulomb's law of electrical interaction) and the resulting AOs are used according to the Pauli principle, one obtains the result that there are *shells* of electrons and, for the atoms Li through Ne, the valence shell — one $2s$ AO and three $2p$ AOs — contains from one to eight electrons. The Pauli principle prevents more than eight electrons from occupying this shell. Similarly, for a transition metal the valence shell of five nd AOs, one $(n+1)s$ AO and three $(n+1)p$ AOs may contain up to 18 electrons.

Why, then, do these rules not *always* work?

There are several factors other than the properties of isolated atoms to be considered when we turn to the much more complicated problem of the distribution of electrons in molecules:

• Although we have used the fact that, close to an atomic core in a molecule, the electron distribution must be similar to that of the separate atom as a justification for the use of (H)AOs in describing molecules, there are very strong electric interactions between other atomic cores and electrons which distort and even disrupt the electron distribution near

---

[2]In rather the same way as the Ptolemaic rules for calculating the motions of the planets are based on *circular* motion, which is one form of motion obtained by Newton's equations for planetary motion.

an atomic core. Atomic orbitals are not like pieces of 'Lego' to be fitted together in a modular way.

- Much more importantly, when electrons are shared between atomic cores in a molecule, their distributions are described by *probability distributions* which, although well-described by combinations of (H)AOs, will not be shared *equally* between those atomic cores. Think of a molecule like HF. The fluorine has a much greater 'pull' for electrons than a hydrogen atom; surely the electron distribution in the bond will be polarised, with the fluorine atom taking the greater share. In fact, a calculation shows that this is the case. The two electrons involved are *shared* between the two atomic cores but, in a rather coarse approximation, the hydrogen atom has only half an electron, while the fluorine takes the other 1.5. The 'number' of electrons around an atom is not such a simple idea when the actual distributions are calculated; this problem is investigated in the next section. The H–F bond is so polarised and the hydrogen atom so poorly populated with electron distribution that it easily attracts other systems with loosely-bound lone pairs of electrons. For example, in solution HF picks up a fluoride anion to form the ion F–H–F$^-$, presenting a conundrum for electron counting: the breaking of the 'duet rule'.

- However, it is obvious that, if the 'pulling power'[3] of the two atoms in a bond are *equal* (or nearly so) then the electrons in a bond will be shared equally between the two atomic cores. Now, carbon is unique among atoms for forming molecules containing bonds *between* carbon atoms. These bonds will have electrons which are (very nearly[4])equally shared between the two atomic cores. Carbon and hydrogen also behave in a very similar way, forming bonds which are not very polarised. Again, calculations verify these simple ideas. So, of all atoms, carbon is, as one might say, predisposed to obedience of the octet rule.

It is therefore clear that the equivalent of there being about eight electrons near a particular atomic core is not necessarily related to just the *number* of bonds to and lone pairs on that atom. One must also know how the electrons are distributed *in space* in the bonds and lone pairs. A bond polarised in one direction might well give an electron distribution near a given atom as great as two (or more) bonds polarised the other

---

[3]Electronegativity is the correct term here.

[4]Carbon atoms in different molecular environments will have slightly different effective electronegativities.

way. And, of course, we should not forget the importance of delocalised π substructures when trying to 'count' the electrons around a given atom.

There are two other concepts widely used in empirical valence theory which are affected by any analysis of the validity of the octet[5] rule:

(1) The idea of resonance in (particularly organic) chemical theories.
(2) The concept of 'oxidation number', which is mainly used by inorganic chemists.

But before continuing it is probably worthwhile to look at the way in which making the very idea of 'counting electrons' is possible within a quantum-mechanical theory.

14.3   Population Analysis

If we concentrate attention on a typical localised σ bond it is possible to give a simple account of how the electrons are distributed and to see how it might be possible to make an approximate count of the number of electrons 'associated' with the two centres involved in the bond.

The way that a localised bond has been described is by forming a bond MO as a linear combination of two HAOs, one on each centre. If the two centres are A and B and the HAOs are ϕ_A and ϕ_B, where ϕ_A is based on centre A and ϕ_B on B. The ϕ_A and ϕ_B are, of course, *functions* of ordinary 3D space, so should really be written $\phi_A(x, y, z)$ and $\phi_B(x, y, z)$, but the dependence on x, y, z is not given to avoid to complicated notation. The MO $\psi(x, y, z)$ (written simply ψ) is therefore given by:

$$\psi = c_A\phi_A + c_B\psi_B, \qquad (14.1)$$

where c_A and c_B are just *numbers* which give the relative amounts of ϕ_A and ϕ_B that, are combined to form ψ. For example, in a non-polar bond, i.e. a bond between two atomic cores which have the same electronegativities, the mixture would be 50% of each HAO: $c_A = c_B$. In general, in a bond which has a greater electron density at the 'A end', the number c_A will be greater than c_B and vice versa.

Now, the electron distribution ('electron density' for short, P, say) due to two electrons occupying this MO to form a localised bond is twice the

---

[5] Or the 18-electron rule.

square of ψ (twice because the MO is doubly-occupied):

$$P(x,y,z) = 2 \times (c_A\phi_A(x,y,z) + c_B\phi_B(x,y,z))^2$$

$$\textit{i.e.} \ \ P = 2(c_A\phi_A + c_B\phi_B)^2$$

$$P = 2c_A^2\phi_A^2 + 2c_B^2\phi_B^2 + 4c_Ac_B\phi_A\phi_B. \tag{14.2}$$

This function gives the *spatial distribution* of electrons due to the localised bond between A and B. The question we want to address to this function is 'how can we get the number of electrons associated with centres A and B?'

First of all, remember that the *function* $\phi_A(x,y,z)$ is a, (H)AO of atom A and $\phi_B(x,y,z)$ is a, (H)AO centred on B, and that the *electron distribution* of any orbital is just the square of that orbital, so we can re-write the above equation in words:

P = (A number)   × (Electron density of HAO on atom A)

    + (Another number) × (Electron density of HAO on atom B)

    + (A third number) × (Overlap of the two HAOs :

                         ϕ_A on A and ϕ_B on B),

where:

- As before, P is the electron density in the bond;
- 'A number' is $2c_A^2$, which is just twice the square of a number and so is itself a number;
- 'Another number' is $2c_B^2$, which is just twice the square of a number and so is itself a number;
- 'A third number' is $4c_Ac_B$, four times the product of the two numbers.
- ϕ_A^2 is 'Electron density of HAO on atom A';
- ϕ_B^2 is 'Electron density of HAO on atom B';
- and recall the definition of the *overlap* of two HAOs, so that $\phi_A\phi_B$ is 'Overlap of the two HAOs'.

Now, it is easy to see that, for example, if ϕ_A^2 is the electron density of HAO ϕ_A and, in the expression for the total electron density in the bond (P), it is multiplied by a number, then that number can be interpreted as the *number of electrons occupying* ϕ_A with a similar interpretation of the number which multiplies ϕ_B^2. We can call these two numbers q_A and q_B respectively. The remainder is the number which multiplies the overlap of the two HAOs: called 'A third number' above. Surely this has to be interpreted as the number of electrons occupying the region of overlap of

ϕ_A and ϕ_B; call it p_{AB}. So we can rename the three numbers in the above expression for P and contract the notation slightly:

$$P = q_A \quad \times \text{ (The electron density of the HAO on A)}$$
$$+ q_B \quad \times \text{ (The electron density of the HAO on B)}$$
$$+ p_{AB} \times \text{ (The overlap of the two HAOs)}$$

The notation for the occupation numbers q_A and q_B has been used for two reasons: the numbers are usually not integers and q is often used in electrostatics for a *charge*. So, a 'number of electrons' is not necessarily a whole number, and it has the dimensions of charge because electrons are charged.

So much is fairly clear, but the natural question to ask is:

Is the *sum* of these two numbers, q_A and q_B, equal to 2 since there are, by assumption, two electrons in the bond?

The answer is, of course, 'not unless the *function* $\phi_A\phi_B$ is *identically zero*' which it cannot be because both the functions ϕ_A and ϕ_B have non-zero values. Thus, we need to know the value of p_{AB}. But this is easy; since $q_A = 2c_A^2$, $q_B = 2c_B^2$ and $p_{AB} = 4c_A c_B$, we have

$$p_{AB} = 2 \times \sqrt{q_A q_B}.$$

Thus, the important conclusions are that the occupation numbers of the two HAOs which are used to form a bond MO:

(1) Do not account for all the electron density of the bond;
(2) Must be augmented by the occupation number of the overlap region: p_{AB}.

If we want to *insist* that we do have some numbers which can be interpreted as the 'numbers of electrons on each centre', we must somehow divide up the number p_{AB} between the two numbers q_A and q_B, which, unfortunately cannot be done in any unique way. There are sensible ways to do this, which are reported by computational methods but which must always be viewed critically.[6] The upshot of this analysis is to emphasise that counting electrons 'associated' with a given atom in a molecule is just a formal

---

[6]For example, it is often the case in transition-metal-containing molecules that the metal atom $4s$ AO may be much larger than the whole molecule. How sensible is it to count the number of electrons in this AO and say that they 'belong' to the metal atom?

exercise and makes no reference to the actual distribution of electrons or to any of the known laws of nature. It is entirely possible that an atom in a molecule which formally satisfies the octet rule has fewer or more than eight electrons or, conversely, one which does not satisfy the octet rule has about eight electrons in its vicinity: it all depends on the sizes of the q_A, q_B and p_{AB}, and the way one divides up p_{AB} and portions out the fractions to A or B. It might seem a reasonable idea to count the number of electrons on each of the atoms in a molecule by, for example, counting the number of electrons which occupy the spherical volume of the isolated atom. The problem here is that the sum of these numbers would always be less than the total number of electrons in the molecule because of all those cusp-shaped volumes, which are the interstices between the spheres.[7] It is also technically very much trickier to compute these numbers.

With these reservations about electron counting in mind, we can now look at the concepts of resonance and oxidation number.

14.4   Resonance and Resonance Hybrids

In chemistry the concept of resonance has grown into a rather formal 'electron-accounting' procedure, whose aim is to generate an electronic structure scheme which:

- Describes the molecule as a 'hybrid' (i.e. a linear combination) of basic units;
- Uses the classical Lewis — dots and crosses — structures as basic units;
- Uses the criterion of minimum formal charge on the atoms to decide amongst the several possible combinations of basic units.

The method is a recipe extracted from the ideas discussed in Section 5.5.1: the valence bond (VB) model. The terminologies 'resonance' and 'resonance hybrid'[8] have a historical significance based on an analogy of the vibrations of solid bodies. These terminologies are very widely used, particularly in organic chemistry.

As usual with empirical rules-of-thumb, this technique is hampered by the fact that it is not based on any laws of nature: there is no reference

---

[7]Try filling up a box with table-tennis balls and it is soon evident that the interstices are quite a large proportion of the volume of the box.

[8]The use of the term 'hybrids' in this context is particularly unfortunate since the same name is also used for hybrid atomic orbitals: HAOs.

to any basic theory. In particular, it is hampered by the fact that the pre-quantum Lewis structures cannot describe the *distribution* of the electrons in a bond or a lone pair. Thus:

- The formal charge on an atom in a molecule which, for a given atom, is simply:

  Net positive charge on the atomic core $-2\times$(number of lone pairs on the atom) $-$ (number of bonds to the atom).

  In a neutral atom the charge on the atomic core is the same as the number of valence electrons.
- Since Lewis structures are used, the assumption has to be made that the electrons in an electron-pair bond are shared equally between the two bonded atoms.

However, since no attempt is made to make the theory a realistic description of the electronic structure,[9] the relative *contributions* of each Lewis structure cannot be found, and the whole process can only be seen as a brave way to avoid any involvement with quantum theory and 'save' the octet idea.[10]

The idea of a *formal* charge *on* an atom in a molecule also appears, by a very different route, in the concept of oxidation number, used mainly by inorganic chemists.

14.5 Oxidation Number

The idea of an oxidation number arose in chemistry in an obvious way from considerations of the oxides of various elements, mainly metal. Since oxygen's most familiar compound is water, where it is combined with just two atoms of (univalent!) hydrogen, and since oxides are more common than hydrides (for metals), one can define a system of empirical valencies of the elements simply by 'twice the number of oxygen atoms with which the element combines in its oxide'. So, for example, the oxide of calcium is CaO and so its valence is two, as exemplified by its other compound

---

[9]In contrast to non-empirical VB theory, which uses the same substructures but backed up by detailed calculations.

[10]In much the same way as Ptolemy used epicycles to 'save' the idea that planets must move in (linear combinations of) circular motions.

$CaC\ell_2$. In these compounds the calcium atom is said to have the oxidation number[11] $+2$.

So much is obvious, even trivial. The difficulties arise, as is usual in empirical concepts, when a simple practically-based rule is elevated by being generalised beyond its utility. In the case of compounds of metals which are *ionic* in the crystal form the oxidation number will be the actual electrical charge carried by the metal atomic ion. So anhydrous crystals of calcium chloride contain the ion Ca^{+2} etc. The history of the inorganic chemistry of metals began as the chemistry of oxides and ionic compounds, and later involved the consideration of metal molecules ('complexes') which had dative and covalent bonds between metal atoms and non-metals.

Unfortunately, the idea of oxidation number was extended to include covalently-bonded atoms (including the non-metals). It has been defined in a way which is consistent with the usage for ionic compounds:

> The oxidation number of an element in a molecule is the charge which the element would have if all the entities to which it is bonded in the molecule were removed in their lowest *closed-shell* state.

So, this still works fine for ionic molecules if the word 'bonded' is taken to mean 'stoichiometric proportion', rather than covalently bonded: the lowest closed-shell state of oxygen is, of course, the dinegative ion with valence structure $2s^2 2p^6$, and so calcium in CaO has oxidation number $+2$ (itself a closed-shell structure). But, if the point of the new definition is generalisation, it should work for molecules in which a metal atom is covalently bonded. Here are some examples:

- Chromium hexacarbonyl $Cr(CO)_6$: the CO *molecule* is a closed-shell system so all six CO molecules may be removed from the Cr atom, taking no electrons away, so that the oxidation number of Cr in this molecule is zero. Straight away, in the simplest possible case, the connection between valence and oxidation number is lost.
- The permanganate ion MnO_4^-: the lowest closed-shell species of oxygen is O^{-2}, so removing the four oxygens takes away eight electrons, leaving the Mn atom with a charge of $+7$ (since the permanganate ion already had a charge of -1). The oxidation number of Mn here is $+7$, a completely unrealistic number.

---

[11]Sometimes called the 'oxidation state'.

- The ammonia molecule NH_3: there are two possibilities for the 'lowest' closed-shell state of hydrogen: H^+ with *no* electrons and H^- with two. Since we are talking about chemistry and not particle physics, the choice has to be the latter: H^-. Removing three negatively-charged hydride anions leaves the nitrogen atom with an oxidation number of $+3$; very large but marginally credible.

- An organic molecule CH_3–CH_3 (ethane): here is an unsurmountable problem. If we take the carbon atom on the left (say) and remove three hydride anions and the closed-shell system CH_3^- on the right, we obtain an oxidation number of $+4$ for the left-hand C atom. Continuing, and removing the three hydride anions from the remaining CH_3^- fragment leaves the right-hand carbon atom with an oxidation number of $+2$. But the two C atoms are completely equivalent in ethane!

It is interesting to compare the definition of the oxidation number as a 'formal charge' with the result of minimising the 'formal charge', using the resonance method of the previous section. An obvious candidate for the comparison is a very common molecular ion, SO_4^{-2}: the sulphate anion, recognised by both organic and inorganic chemists.

- Without going through all the details, the result of applying the resonance recipe to the bonding in the sulphate anion is that it is a resonance hybrid, consisting mainly of equal amounts of all the possible ways of joining two of the oxygen atoms to the central sulphur by single bonds, and the other two oxygens are attached by double bonds, indicated schematically as:

$$
\begin{array}{ccc}
 & O^- & \\
 & | & \\
O \!\!=\!\!\!=\!\! & S & =\!\!\!=\!\! O \\
 & | & \\
 & O^- & \\
\end{array}
$$

Any one of these structures obviously does not have the symmetry of the actual sulphate anion, but an equal mixture of all six possible equivalent structures does. Using the definition of formal charge on page 245 gives the sulphur a charge of *zero* and each oxygen a charge of $-1/2$.

- The oxidation number of the sulphur atom is found by removing all four of the oxygen atoms as O^{-2} anions, and taking account of the overall charge of sulphate of -2; the oxidation number of sulphur is *plus six*.

The difference between these two 'formal charges' is not trivial! So long as the golden rule:

> Organic and inorganic chemists shall not converse with each other about their rules

is obeyed, there will be no difficulty. But if the behaviour of real electrons in real molecules is considered, there is bound to be trouble.

14.6   Summary for Number Rules

It has been the main thrust of the material presented in this work that any theory of the electronic structure of molecules *must* be shown to be consistent with the known laws which govern the energies and distributions of systems of charged particles in mutual interaction. It was stressed at the very outset that the behaviour of electrons in molecules cannot be *reduced* to these more basic laws, any more than the mating display of a peacock can be *reduced* to the laws of chemistry which, however, *must* be obeyed during that display. All biological processes must obey the laws of chemistry and all chemical properties must obey the laws of physics but, in neither case, can the 'higher' process be simply *derived* from the lower.

In the early stages of any branch of science there are bound to be empirical rules-of-thumb which help to correlate the raw data and rationalise the known facts. Sometimes these rules turn out to have a genuine basis in the underlying theory describing the phenomena, but sometimes they do not. There is always a very understandable desire to modify and extend a familiar 'explanation' in the face of new evidence which apparently decisively refutes the current theory. However, when this turns into attempts to evade a newer and more comprehensive theory by ignoring that theory, science is being debased. There is nothing wrong with using some of these rules as mnemonics — tricks to remember some facts — in much the same way as we use a simple rhyme to remember the number of days in each month, but they cannot be a substitute for understanding.

The behaviour of matter at the chemical level is very complicated and subtle. It is not reasonable that the electronic structure of molecules can be described or even summarised by a few rules (whether or not they are based on quantum mechanics). If the systems of charged particles interacting by Coulombs's law, Schrödinger's mechanics and the Pauli principle can generate tigers, computers, stick insects and steam engines, it is to be expected that a very subtle theory is needed to try to get to grips with even a very specialised part of these structures: individual molecules.

What, then, remains of the rules? It is increasingly clear that the major factor which tends to fix the *upper limit* to the number of bonds which a given atom may form — the *maximum* valency of the atom — is not the number of valence electrons which are associated with that atom but *the number of electrons which that atom is capable of sharing with other atoms.* When this criterion is applied we can see a much more rational underlying trend. What is more, the atom which obeys the octet rule most consistently is the carbon atom, for which the octet rule and the electron-sharing criterion give exactly the same answer, *because having four electrons to share in chemical bonds without any lone pairs present automatically means that the octet rule is obeyed.* The 'number of electrons which an atom is capable of sharing' means just that, and not 'the number of valence electrons in the separate atom'. Nor does it imply anything about the *distribution* of those shared electrons; they may be equally shared or (very) unequally shared. So the definition includes dative bonds and highly polar bonds, as well as the usual homopolar covalent bond.

Thus, for example:

- The boron atom, with only three valence electrons, is capable of sharing eight electrons when forming the molecule BH_3CO; sharing three of its own electrons and three hydrogen electrons plus two electrons from the lone pair of the CO molecule, as we saw in Chapter 10.
- Atoms which exhibit 'hypervalence' are capable of sharing fewer than their number of valence electrons (S in H_2S and SF_4) in some molecules, as well as all their valence electrons in other molecules (S in SF_6), phenomena discussed in Chapter 13.

14.7 Assignment for Chapter 14

(1) Write an account of the octet rule and the 18-electron rule and their uses as an introduction to chemical valence theory, and indicate how you would use them to introduce the idea of bonding to someone who knows no chemistry.
(2) Now do the same thing for someone who has some knowledge of quantum theory; in other words, justify these rules as far as you can.

Chapter 15

Mavericks and Other Lawbreakers

No-one should ever imagine that the behaviour of nature can be explained by means of a few simple rules. All science is 'abstract', in the sense that it can only deal with typical cases, and nature provides us with constant surprises with brand new 'typical cases' all the time as well as exceptions to the typical cases. The most important exception to the 'electron-pair' model is, of course, the delocalised π system of Chapter 11. Here are a few which we did not think about earlier.

Contents

15.1   Exceptions to the Rules

By far the majority of the electronic structures of what one might call 'familiar' molecules can be understood using the substructures we have been discussing: the electron-pair structures (localised bonds and lone pairs) and the delocalised π systems of conjugated molecules. There are, however, other structures which, while not as common as these examples, are sufficiently common to be worth a short account. By far the richest source for the discovery of new electronic substructures of molecules is the electronic structure of transition-metal compounds, particularly those containing the metals of the second and third rows of the transition series. These await investigation and so cannot be described yet. A guess at one possibility has been made in Section 15.3.

In situations where the qualitative structure of a molecule or type of molecule is not known, the standard method of quantum chemistry can be used to obtain at least some quantitative information about the overall electron densities and energetics of any system. Recall the general strategy which was worked out very early on in our investigations, summarised in Section 3.2.2.1:

- Molecules are composed of charged particles (electrons and nuclei) and must, therefore, be subject to the laws of physics: Coulomb's law of attraction and repulsion, Schrödinger's (quantum) mechanics and the Pauli principle.
- Since, from one point of view, molecules are made of atoms,[1] we have seen how atomic orbitals (AOs) are suitable building blocks for a chemist's approach to solving the equations which give the distribution and energies of electrons in molecules.

We can therefore *always* get quantitative information about molecular electronic structure by using a 'brute force' approach.

> We can use all the AOs of all of the atoms of a molecule to expand all the MOs of that molecule, not giving any further thought to the possible substructures into which the overall electronic structure can be usefully understood in a qualitative way.

---

[1]Remember the discussion in Section 1.2.

For example, in Chapter 7.3 the AOs of the carbon, oxygen and two hydrogen atoms of methanal were used in the following way:

- Take the 'raw' AOs and form polarised HAOs to describe the localised *sigma* bonds and lone pairs.
- Form linear combinations of these HAOs to describe the MOs of the C–O and C–H σ bonds, and leave each of the oxygen lone pairs described by an HAO.
- Use the π-type $2p$ AOs on carbon and oxygen to form an MO describing the π bond between those two atoms.

We could, however, have simply taken the ten AOs of the constituent atoms of the molecule[2] and used *all* of them to expand six MOs to accommodate the 12 electrons of methanal, i.e. each MO would be composed (at least in principle) by a linear combination of all ten AOs. This would generate a set of MOs which would have the correct total electron density but would yield hardly any useful information about the chemical structure of the molecule. This information can only be extracted if we know what that qualitative structure is and transform the MOs into something which reflects that structure. However, it does work for the gross electronic structure and can always be used as a first approximation or as a last resort.

There are, of course, cases where the actual *chemistry* of the system suggests new types of *qualitative* substructures, and it is one or two of these which are looked at here, just to emphasise that the bulk of the discussion so far, although covering most (organic) molecules, does not always apply to the most interesting cases!

15.2 Boron Hydrides and Bridges

The chemistry and structures of the hydrides — 'hydrides' notice, not hydride — of boron were only discovered and partially understood in the middle of the 20th century, in spite of the fact that boron had been known for many years before. They provided a complete new branch of inorganic chemistry with interesting parallels to the chemistry of carbon.

---

[2]$2s, 2p_x, 2p_y, 2p_z$ of carbon $2s, 2p_x, 2p_y, 2p_z$ of oxygen, and $1s$ of both hydrogens.

15.2.1   *The expected compound:* **BH₃**

The simplest hydride of boron might reasonably be thought to be BH_3, in which:

- The atomic core of boron is the nucleus plus the two $1s$ electrons.
- There are three equivalent B–H σ bonds, each described by an MO composed of an HAO on boron and the $1s$ AO of a hydrogen atom. This would neatly account for all eight electrons of the suggested molecule.

This molecule would be planar, with each H–B–H angle being 120 degrees.[3] In fact, the BH_3 molecule does exist[4] but is extremely reactive, and it is worth a short digression on why this should be so.

The three B–H bonds in the molecule repel each other, leading to its symmetrical planar shape, but these six electrons are the total number of valence electrons. This means that the regions above and below the plane of the molecule, while not devoid of electron density, because the electron distributions of the B–H bonds are cylindrically-shaped, are relatively thinly populated by electron density. That is, above and below the plane, the electrostatic field of the boron atomic core is not very effectively shielded by electron density. In the similar molecules CH_4, NH_3 and H_2O, the atomic cores are shielded in all directions by the electron density of either bonds or lone pairs.

> There simply are not enough electrons in BH_3, either to form three B–H bonds or provide effective shielding of the atomic core from all directions: this molecule is colloquially said to be 'electron deficient'. This is just a dramatic way of saying, in a short-hand way, that there are regions of space sparsely populated by electrons, and certainly not that BH_3 has fewer electrons than it should have to make it electrically neutral. There are eight positive charges on the nuclei of the molecule and eight electrons.

Now, we know what happens when there are regions of space near atomic cores which are not well shielded: they attract negatively-charged electrons.

---

[3] Check this by using the method of Section 6.3.
[4] As usual, 'exist' is used here to mean 'not falling apart spontaneously'.

So we expect that the BH_3 molecule would be a target for reaction with molecules containing lone pairs of electrons and, sure enough, the NH_3 and CO molecules do form dative bonds with the structures:

$$OC \longrightarrow BH_3 \quad \text{and} \quad H_3N \longrightarrow BH_3.$$

If that were the extent of the chemistry of boron hydride it would be unsurprising and explicable in terms of the structures we have met already; but it isn't.

15.2.2 *The compounds which are found*

The simplest hydride of boron which is stable and (relatively) unreactive is the 'dimer' of the expected compound B_2H_6. But what is its *structure* and what are the electronic substructures which it contains? Further research showed an embarrassing richness and diversity of hydrides of boron and, when the geometrical arrangement of the boron and hydrogen atoms was finally found, an equally embarrassing set of structures which contained, among other things:

- Hydrogen atoms which had *two* boron atoms as nearest neighbours, both close enough to be counted as 'bonded' to the hydrogen atom. Hydrogen atoms with an apparent valency of *two*!
- Boron atoms with two other boron atoms as equidistant near-neighbours *plus* two near-neighbour hydrogen atoms. Boron with a valency of four but with no dative bonding partner as in BH_3CO, so where are the electrons coming from to form all these 'bonds'?

There were a number of false starts at the explanation of these phenomena, but it soon became obvious that what was seen in the boron hydrides was a new electronic substructure which is similar in similar molecules, and is just as insensitive to its environment as the now familiar two-atom electron-pair bonds.

15.2.3 *Bridged, three-centre bonds*

The actual molecular geometry of the simplest boron hydride (B_2H_6) is similar to that of ethene, but with two additional hydrogen nuclei on either side of the plane formed by the two boron and four hydrogen atoms, symmetrically above and below the mid-point of the boron–boron distance:

However, there are two distinct B–H distances: four in the ethene-like plane and four in the perpendicular plane. There are clearly two types of B–H 'bond'.

Similarly, there is a stable molecule, B_5H_9, which contains both of the novel types of structures: 'doubly-valent' hydrogen and 'quadruply-valent' boron.

> These structures were both seen to be examples of a new molecular electronic substructure: the **three-atom two-electron bond**. In this substructure the central atom of the three is not in line with the other two, but is said to be a **bridge**[5] joining those two.

What we need to do now is to bring this novel, law-breaking structure into the fold of known environment-insensitive electronic substructures; in a phrase, to see how it fits in with our knowledge of the laws of nature outlined in Section 15.1, which all electrons must obey.

First things first: how do these new electronic substructures fit — if, indeed, they fit at all — into the methods and models used so far? In particular, can we form a doubly-occupied molecular orbital which describes the three-centre bond? Since there are three centres involved and, as has been repeatedly stressed, the electron density near any nucleus must be very similar to that of the corresponding atom, there has to be contributions to any MO from (at least) *three* (H)AOs. The obvious thing to try is four standard B–H bond MOs formed from a σ-type HAO on the boron atom and the $1s$ AO of the hydrogen atom. Then, presumably, each of the two B–H–B bridge structures can be described by an MO composed of a

---

[5]The idea here is of the old 'hump-backed' bridge, not a flat suspension bridge!

similar 'σ-type' HAO on each boron atom *and* the $1s$ AO of the bridge hydrogen.

This proves to work very well, and the two types of bond (B–H and B–H–B) are illustrated in the contour diagrams below:

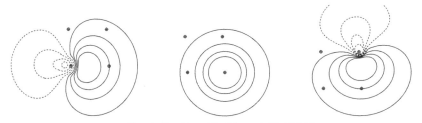

Contributing B, H and B HAOs

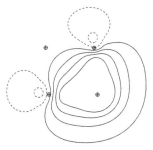

B–H–B bridge MO

The plane of the paper contains the two boron atoms and the two bridge hydrogens; the four 'terminal' hydrogen atoms cannot be seen as they are either above or below this plane. Slice contour diagrams are used, since the single-contour 3D diagrams are much less clear. The four terminal B–H bonds are not shown, since they are qualitatively similar to the σ C–H bonds shown elsewhere in the text. The H–B–H angles between adjacent terminal B–H bonds is very close to those of the similar ones in ethene, about 121 degrees, while the two B–H–B bridge bonds are much smaller: about 86 degrees.

Similar calculations can be carried through for the $B_3H_8^-$ anion, which has terminal and bridge bonds between boron and hydrogen *plus* a more conventional-looking structure with a single bond between two borons in place of the bridging structure. The whole system forms an isoceles triangle of boron atoms:

The structure of the bridge bond between three boron atoms, which is found in the icosahedral structures of boron, is qualitatively similar to the B–H–B structure but involves three boron HAOs; one of them replacing the hydrogen $1s$ AO in the resulting MO.

Bridge (three-centre) structures are also common in transition-metal chemistry, and a huge variety of new substructures have been found which await a satisfactory qualitative description.

15.3   Other Three-Centre Bonds?

As we shall see in Chapter 16, if we are looking for unusual electronic substructures the first place to look is molecules containing transition-metal atoms. There are many three-centre structures found in the chemistry of these metals. In particular, the CO molecule, which is always ready to share the electrons in its carbon-atom lone pair, readily forms bridges between pairs of iron atoms. In these structures, unlike the hydrogen atoms in the boron hydrides, the CO molecule provides two electrons to the three-centre bridge.

In the comparison of bonding in SO_4^{2-}, MnO_4^- and OsO_4 (Section 13.5.1) a possibly significant difference between the σ-bond MO in SO_4^{2-} and the σ-bond MO in MnO_4^- and OsO_4 was noted. Each σ-bond MO involving Mn or Os has, as well as the familiar lobe indicating concentration of electrons between the two atomic cores, a large lobe pointing *away* from the bond region. If we look at the forms of the individual HAOs involved in the metal-to-oxygen σ bond there is a striking difference between these and the corresponding sulphur HAO, which has a familiar 'sp^3' look:

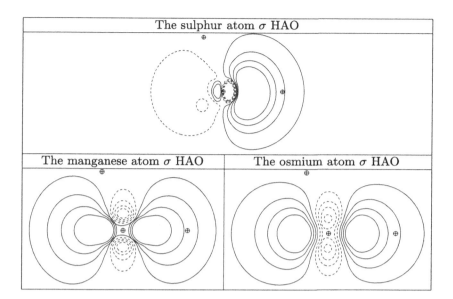

The metal HAO is nearly symmetrical, with similar lobes along the bond and in the opposite direction. In the tetrahedral molecules considered in Section 13.5.1, these 'spare lobes' point in directions where there are no other atoms. That is, the electron density in these lobes is not shared and, presumably, does not contribute to the bonding.

Now, consider the more common case of a square planar transition-metal molecule of the form ML_4. If we take a slice through such a molecule in a plane containing the four L ligands (atoms, say, for convenience) it looks like:

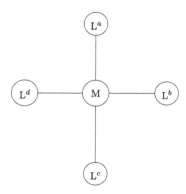

(where the ligands have been labelled L^a, L^b, L^c and L^d for convenience).
By analogy with σ-bond MOs in the tetrahedral case, we might expect a σ
bond between M and each of L^a, L^b, L^c and L^d. Taking the case of M–L^a
it is easy to see that there will be a bond MO formed by combination of
an HAO from M and one from L. Usually, M is the transition-metal atom
and L a typical element. So, just like the tetrahedral case, there will be a
lobe on the M atom pointing away from the bond. But that is precisely the
direction of the ligand L^c. So, looking at the molecule with this in mind, it
is clear that the ligands L^a and L^c have almost 'equal claims' to the HAO,
which we began by considering as involved in a *localised two-electron* bond
between M and L^a! The metal HAO has lobes pointing at each of L^a and
L^c which are equidistant from it. If these two lobes are very similar there
may be no way of choosing which ligand to form a localised two-electron
bond with.

This dilemma is recognisable as being of exactly the same type as the
one we met in our discussion of conjugated π systems in Chapter 11, where
there was no unique way of deciding on 'bonding partners' for the π AOs.
The solution to that dilemma was the theory of a *delocalised* π substructure
for these molecules: the conjugated π structure. It would seem at least likely
that the situation here also demands a re-think of the conventional localised
electron-pair bonding scheme, but this time for a *σ-bonded system*.

We cannot just spin theories out of our heads and expect them to be
true because of some analogies with other structures. The actual properties
of these molecules must be looked at (as we did for delocalised π systems)
to help to make a choice. What *are* the differences?

- Firstly, and apparently most obviously, one would think that, since there
  are two electrons per bond in the traditional model and only a net one
  per bond in the three-centre model, the bond in the latter case would
  be weaker than, for example, an M–C≡O in a tetrahedral molecule.
- Secondly, it is found in the chemistry of square–planar compounds that,
  if a ligand substitution is made at one site (say we replace L^a by X in
  our diagram on page 259) this has an effect on the properties of the
  other three M–L bond properties. The effect is often strongest in the so-
  called *trans* position, i.e. *diagonally* opposite the position of X. In our
  case this would be the M–L^c bond. This looks more positive for, if the
  three atoms X, M and L^c are involved in a (σ) delocalised structure, the
  transmission of electron density across the three atoms will be easier than
  the influence exerted on each other by two independent electron-pair

bonds. Also, there is no apparent reason why two 'ordinary' electron-pair bonds should interact more strongly in the *trans* position than if they are nearer in nearest-neighbour (*cis*) positions. If two ligand atoms in the *trans* relationship were involved in a common bond with the metal atom, it is clearly possible for (e.g.) a very electronegative ligand to weaken the *trans* ligand by pulling the electrons away from it.

• Finally, since it is possible to form three MOs from three HAOs, there is always the possibility that there are two doubly-occupied MOs involved in the three atoms in the *trans* positions. If this were to be the case, the second MO would be of the non-bonding type, with major contributions from the two ligand HAOs and very little (none, if the molecule were completely symmetrical) contribution from the metal-atom HAO.

Of course, the final analysis must come from detailed calculations and careful comparisons with experimental data.

15.4 Metals and Crystals

Metals and crystals are solids and they do not have the sort of electronic structures which chemists are normally concerned with, but they are composed of charged particles (electrons and nuclei) so they should, perhaps, have some connection with the electronic structures of molecules.

15.4.1 *Metals*

The characteristic property of the electronic structure of metals is that there is a substructure which is the source of their electrical conductivity. Conductivity was encountered in Chapter 11, where the π-electron structure of graphite was briefly mentioned. What is involved in graphite and, presumably, in metals is a *macroscopic* delocalised electronic substructure. The application of a source of electrons to one end of a piece of metal — a copper wire, say — provides electrons at arbitrarily large distance along the metal: hundreds of miles in the case of electrons supplied by a power station. The question is: can the (H)AO-based model provide an explanation of this phenomenon?

A physicist, asked about the possibility of describing the structure of the 'conduction band' (the substructure which gives metals their conductivity) in terms of HAOs, would give a very brusque answer; probably something like 'certainly not'. The key thing about graphite and polyethyne is that the 2D (or 1D) conductivity of these 'molecules' is provided by delocalised

π substructures, while the whole molecule is held together mainly by the σ bonds. This rather limits the conductivity to be described in this way to a maximum of two dimensions, while metals will conduct electricity in *any* direction; in any combination of all three independent directions in the solid.

The 'conduction electrons' in a metal almost behave as if they are free to move about the whole structure and can effectively disregard the attractions of the metal-atom atomic cores, so a physicist will *start from* this idea rather than trying to arrive at it using (H)AOs. Of course, the solutions of the quantum-mechanical equations for free electrons are very different from the familiar AOs, which describe tightly-bound electrons. They are the so-called 'plane waves': sine and cosine functions or their complex combinations. So, in this case, our technique of using AOs is completely inappropriate for this important substructure. However, the main idea of environment-insensitive electronic substructures is still valid. The only point is that the way we have implemented this idea in chemistry is of little use for this substructure. The situation is, in one sense, similar to the molecular case. To describe a particular electronic substructure one must use functions which are known to be adapted to the description of that substructure: (H)AOs for localised structures in molecules and plane waves for (nearly) free electrons.

On the other hand, if one is interested not so much in electrical conductivity but in what it is that holds the atoms together to form a solid metal, then the methods of molecular electronic structure *do* have a part to play, since the conduction electrons only make a minor contribution to the cohesive energy. There are the usual atomic cores on each metal atom, and the valence-shell AOs can be combined together to make bonding MOs in much the same way as the (H)AOs of molecules. So, in order to describe the *overall* electronic structure of a metallic solid, one would use:

- Inner shell AOs to describe the atomic cores of each atom.
- Linear combinations of (H)AOs to describe the bonding (cohesive property) of the metal crystal structure; what physicists call 'tight binding', since these electrons are not free to move as the conduction electrons are.
- Linear combinations of plane waves to give an adequate description of the conductivity of the metal.

In practice, since the conduction electrons are not actually free but do experience a pull to the atomic sites, the conduction electrons are often described by linear combinations of both AOs *and* plane waves.

15.4.2 Crystals

For our purposes crystals can be thought of as falling into four types:

(1) Metals.
(2) Ionic crystals: salts of various kinds, common salt being the most familiar example.
(3) Crystals of the non-metallic elements: carbon, silicon, boron etc.
(4) Molecular crystals: the solid state of any pure compound: benzene, sugars etc.

These categories are not exclusive, of course. There are molecular ionic crystals, for example, and the basic units of phosphorus and sulphur are molecules of linked atoms of the element.

We have already said a few words about metals and, although it is an over-simplification, the structure and the forces that form ionic crystals are fairly easy to understand: the mutual attraction of opposite electrical charges and the packing of objects of different shapes and sizes are the main factors. By and large, the electronic structure of crystals of the familiar non-metallic elements can be understood as simply the electronic structure of very large molecules. As we have seen, graphite can be considered to be very large conjugated molecules, akin to the naphthalene and anthracene molecules, and diamond is a 3D structure of tetrahedral C–C σ bonds, while boron is based on an icosahedral structure of B–B bonds, and B–B–B bridging σ-type bonds. The atoms in these crystals are covalently bonded to each other.

Molecular crystals differ from the other three categories in one very obvious way: if they are not ionic they tend to have melting points which are very much lower than metals, ionic or covalently-bonded crystals. They also are mechanically much weaker (easy to crush). This means, of course, that the forces which bind the molecules together are very much weaker than ionic or covalent bonds.

In discussing the shapes of polyatomic molecules, attention has been focused on the *repulsions* between distributions of electrons within those molecules. But in our very early thinking about the mechanism of bond formation we had to consider the fact that atoms, which are surrounded by electrons, still exert mutual attractions. Now, in thinking about the mutual interactions of molecules, we must bear in mind that, since *all* substances, if cooled and pressurised sufficiently, will condense into liquid and then solid form. *All* molecules must attract one another to a greater or lesser degree. What is the source of this attraction between systems in which the outer structure is a density of particles *with the same charge* (electrons)? There are several possibilities, among which are:

- If, for example, the electron distribution of the molecules is not spherically symmetric — and no molecules are *exactly* spherical — then the molecule will have a dipole moment,[6] and there is a force exerted between dipoles. If a dipole is represented crudely as $+ \longrightarrow -$ then two such dipoles will tend to line up head-to-tail in line or alongside or some intermediate way, depending on the shape of each molecule.
- In addition to the force exerted between the dipoles as they are, the dipole of each molecule is caused by the layout of the (positively-charged) nuclei and the *distribution* of the surrounding electrons. Any other charge or dipole approaching this outer distribution will exert a force on the electrons and polarise them: change their distribution. Equally, the other approaching dipole will be changed in the same way, so that the net forces between the two molecules will be *greater* than if they were 'rigid' dipoles unable to be polarised.
- Finally, even molecules which have no dipole can be distorted by collisions etc. and each induce dipoles in each other, and these two mutually induced dipoles will exert a force between the two molecules. Thus, even spherically-symmetric electron distributions will exert attractive forces on each other.

It is to be expected on the basis of this reasoning that all molecules, if the temperature is low enough to allow these relatively weak interactions not to be overcome by the energy of collisions, will condense through their mutual interactions to form liquids and, ultimately, solid crystals. The more polar the molecule, the more easily will the compound form a condensed phase, with the less polar atoms and molecules (rare gases, methane) solidifying at the lowest temperatures.

15.5   The Hydrogen Bond

One particularly strong form of intermolecular interaction, which is weaker than ionic or covalent forces but stronger than the forces discussed above,

---

[6]Or a quadrupole or higher moment, but for the present purposes, the use of a dipole is the most familiar.

is the so-called 'hydrogen bond', which explains, among other things, why water condenses and solidifies much more readily than other molecules of similar size.

When a σ bond (X–H, say) is formed between hydrogen and another atom X, the electron distribution in that bond is governed by the relative electron-attracting power (electronegativity) of the H atom and the atomic core of atom X. Now,

(1) If X is very electronegative, most of the electron distribution will be pulled towards X away from H, leaving the atomic core (the positively-charged nucleus) of the H atom relatively exposed;

(2) We have already seen that an atomic core which is exposed in this way is attractive to electrons; this is the classic way in which a dative bond is formed;

(3) Hydrogen's nucleus is not, however, capable of attracting two electrons to form a dative bond, but the attractive force is still very much present.

What actually happens is that the hydrogen end of an X–H bond in which X is highly electronegative (O and F on the right of the periodic table) will pull lone pairs of electrons towards itself with a force which is weaker than a (datively-formed) covalent bond, but much stronger than most other intermolecular forces.

In water itself there are two polar O–H bonds and (fortunately, as it were) two lone pairs, so the water molecules can form these hydrogen bonds with one or two partners. In solid water — ice — the water molecules form a structure in which each O atom of each water molecule is surrounded by four hydrogen atoms: the two to which it is covalently bound, and two others from other molecules via the hydrogen bonds between their H atoms and its lone pairs. The structure is actually tetrahedral, like the structure of diamond. Some indication of the relative strengths of a covalent bond and the hydrogen bond is obtained by comparing the properties — hardness and melting point — of ice and diamond!

15.6 Lawbreakers?

These few examples of electronic structures which do not fit into the schemes developed earlier do not bring the whole idea crashing down. The main guiding plan introduced in Chapter 1 is still very much alive:

> In order to understand the behaviour of any complex struc-
> ture, one has to use knowledge gathered from experience and
> observations to break down the complicated whole into parts
> which are, as far as possible, independent. These parts —
> which are *relatively* insensitive to their environment — may
> then be examined by appropriate techniques.

The new ideas of three-centre bonds, the hydrogen bond and the new
structures found in crystals and metals can all be treated in this way
by extending the methods developed to deal with more 'conventional'
molecules. When this is done, it is always the case that the three main
laws which govern the energies and distributions of the component charged
particles, of which all matter is composed, can be applied qualitatively and
quantitatively.

No laws are broken; new ones are discovered.

15.7   Assignment for Chapter 15

Inorganic chemistry — particularly transition-metal chemistry — is a rich
source of 'maverick' compounds; that is, molecules containing electronic
substructures which do not belong to the categories we have looked at.

Find some examples from your notes or your text-books and see if you
can find molecules containing 'bridging' structures similar to those in the
boranes. Is the very general idea of environment-insensitive electronic substructures using HAOs any help in trying to understand these structures,
or is it necessary to have a completely different approach to a qualitative
understanding of the bonding in these molecules?

This is difficult; a problem for discussion with your tutors well beyond
what has been introduced in this text!

Chapter 16

The Transition Elements

The 'transition' metals are so called because they interrupt the sedate progression in the periodic table through groups of just eight elements. They are a different type of atom from both the non-metals and the metals of the rest of the periodic table. In the case of the typical[1] elements there is an abrupt change in chemical properties with atomic number, i.e. each additional electron. In the transition elements this sudden change in properties is replaced by a gradual change from metal to metal, reflecting their different type of atomic electronic structure. Naturally, this means that molecules containing the transition metals will exhibit some properties which we have not met so far and the relationship among the properties of these elements will be different from those of the typical elements. It is important to emphasise, however, that, at the moment, there is no coherent qualitative theory of the bonding in transition-metal molecules.

Contents

---

[1] The 'typical' elements in the periodic table are those found in the same columns as the elements lithium to neon.

16.1   The Background

The terminology used in describing the structure and chemistry of transition-metal compounds is different from the sort of description we have used so far: the most obvious example is the way in which molecules containing transition metals are called 'complexes' rather than 'molecules', usually without explanation. This choice of nomenclature is due to the history of the chemistry of these metals, rather than anything *fundamentally* different about the chemical bonding and electronic structures involved. As we have seen in Chapter 13, what *is* very common in transition-metal-containing molecules is also known in the chemistry of the heavier 'typical elements':

- Higher valence than the light typical elements; it is very common, for example, for an atom like iron or chromium to be bonded to six other atoms.
- Variable valency; again, the heavier 'typical elements' may form compounds in which a given atom may have different numbers of bonds.

However, these features are very much more common in the transition metals. Indeed, the ability of a transition-metal atom to bond to four, five, six and sometimes more atoms is the typical property in their chemistry. It is also the case that the *type* of bond formation among these metal atoms is, typically, dative — making a covalent bond by sharing two electrons from just one of the bonded partners — rather than the 'standard' covalent bond formation involving the sharing of one electron each in an electron-pair bond, which is more typical of the cases discussed in Chapter 13.

It seems obvious that we should look for an explanation of these new features in terms of the way in which the electronic structure of the transition-metal atoms differs from that of the 'typical' elements.[2]

16.2 Transition Metals: Effects of 'd' Electrons

The familiar diagram of the energy levels available to electrons in atoms of the first three rows of the periodic table — hydrogen to argon — and the rules of use of this diagram give an excellent indication of the *periodicity* of the chemical properties of the atoms:

- Because the ns/np set of levels is filled by eight electrons, similar properties recur for an atom of atomic number Z and one of atomic number $Z + 8$, like Be and Mg or F and Cℓ.
- This convenient scheme is completely disrupted at $n = 3$, since a new set of five levels capable of holding ten electrons appears: the $3d$ set.
- These $3d$ levels are similar in energy to the $4s$ and $4p$ levels.
- Recall the general pattern of the atomic energy levels: they depend on $-1/n^2$ so they get closer and closer together with increasing n *and* the nd is higher than the np which, in turn, is higher than the ns.
- This means, of course, that, sooner or later, the ns level will 'collide' with the $(n-1)d$ level.

In fact, this happens sooner rather than later and the $3d$ and $4s$ levels are very close in energy.

The effect of this is to create a kind of 'double periodicity' in the table:

- When it is the turn of the nd levels to be filled they all have a similar 'outer' electronic structure of two electrons in the $(n + 1)s$ level, so there will be a tendency for them all to have similar chemical properties. They only differ by the number of electrons in the nd levels and these will accommodate *ten* electrons. So we expect the 'main' periodicity of the table to be interrupted by a set of ten metals with similar chemical properties to each other but different from those of the typical elements. When these nd levels are all filled, the 'main' periodicity will be resumed. So, for $n = 3$ the periodicity is interrupted after calcium (atomic number 20) and resumed again at gallium (atomic number 31).

---

[2]It should be said that the 'rare earth' or lanthanoid metals (cerium to lutetium; atomic numbers 58 to 71) exhibit similar properties, but they will not be discussed here.

- But, *within* the series of similar metals, one would naturally expect greater similarities between atoms with the same number of '*d* electrons' than there is between atoms with different numbers of *d* electrons. So, for example, cobalt, rhodium and iridium (each with seven *d* electrons) have chemical properties which are more similar to each other than those of cobalt are to titanium or tungsten to gold.

It is these new structures which generate the differences between the properties of the transition metals and those of the typical elements.[3]

16.3   'Screening' in the Electronic Structure of Atoms

A glance at the familiar diagram of the energy levels available to electrons in metals shows that the *energies* of (e.g.) the 3*d* and 4*s* AOs are very similar, but what this simple diagram cannot show is the very important *differences* between the AOs themselves even though their *energies* are very similar. If we plot contour diagrams to the same scale of the 4*s* AO and one of the 3*d* AOs of one of the atoms scandium through copper, the main differences quickly become very obvious:

- Most obviously, the *shapes* of the two AOs are very different; the 4*s* AO is spherically symmetrical while the 3*d* AO has the familiar 'double dumbbell' shape.
- However, the relative 'sizes' of the two AOs are equally significant for the chemical properties of these atoms. The 3*d* AO is much more compact than the 4*s* AO which is, by far, the most diffuse of the two.

If we now think about what this means for the *chemical* properties of these elements we can begin to make sense of their high and multiple valences. In what follows, the 4*s* and 3*d* AOs will also do duty for the general case: ns and $(n-1)d$ for $n = 4$ and 5.

In the sense of both energy and size the 4*s* AO containing two electrons (or one in the case of chromium) is the outer electronic structure of these atoms and these electrons are the ones most easily polarised, shared or even lost altogether during molecule formation. However, in the transition metals, there is the possibility of *adding to* the number of 3*d* electrons

---

[3]It is worth noting that, among the 'natural' elements (hydrogen to uranium) there are 44 'typical' elements and 48 transition elements: another historical legacy.

which lie 'inside' the $4s$ AO.[4] Recall that electrons interact via Coulomb's law — they repel each other — and there is a very important consequence of this law which has been known since the early days of electrostatics:

> A charged sphere exerts no net force on any charges inside that sphere.

Now, an electron in a $4s$ AO is not a charged sphere, but it does have a spherically-symmetric charge *distribution* and the consequence of this is that the $3d$ electrons which lie inside the $4s$ charge distribution are not strongly affected by the presence of the $4s$ electrons. Also, again because the $3d$ AOs lie inside the $4s$ AO, they are closer to the nucleus (on the average; remember it is a charge distribution we are talking about) than is usual for valence electrons so they experience a stronger attractive pull. The short-hand way of saying all this is that:

> The electrons which occupy the $3d$ AOs are less *screened* from the electrostatic pull of the nucleus than is usual for valence electrons.

So, suppose that we take a particular example, nickel (Ni) say, and do some mental chemistry with this atom.

- Nickel's outer electronic structure is $3d^8 4s^2$ and, from what we have said above, will share or lose its two $4s$ electrons most readily. We can imagine that two chlorine atoms will each pull off one of these electrons to form an ionic compound $NiC\ell_2$.
- We would expect $NiC\ell_2$ to dissolve in water and dissociate into ions as follows:

$$NiC\ell_2 \longrightarrow Ni^{2+} + 2C\ell^-,$$

where the nickel atom now has the outer electronic structure $3d^8$ only.
- But a full set of $3d$ AOs contains ten electrons, and so the nickel atom will now be able to exert a strong electrostatic attraction on other electrons precisely because these AOs are less screened than other valence

---

[4]Remember here that the (squares of) the AOs give the *distribution* of the electrons which are spread out in space, so we must be careful to understand what is meant, in this context, by 'inside'. It is used to mean that most of the electron distribution of a $3d$ electron lies closer to the nucleus of the atom than the bulk of the distribution of a $4s$ electron.

electrons. The nickel ion in aqueous solution will, therefore, be 'looking' for any loosely-bound electrons in its immediate environment, which is, of course, water molecules.

- Where are there such electrons? The most obvious candidates are the lone pairs of the solvent water molecules, of which there are plenty around. All that has to happen for the nickel ion to satisfy its need for more electrons is that water molecules should approach it in the correct orientation: with their lone pairs closest to the atom rather than their hydrogen atoms.

The nickel ion is, of course, carrying a positive charge of two units, so this fact alone will tend to make the water molecules turn their most negatively-charged part towards the ion as they collide with it in solution. So the stage is set for the doubly-charged nickel ion to surround itself with water molecules with their lone-pair electrons close by.[5]

This kind of behaviour is very common in the properties of the transition metals. What is also noticed when transition-metal salts are dissolved in water (they dissolve very readily) is that very often the act of dissolving is associated with *changes in colour*: the solution is a different colour from the dry salt. This certainly contrasts with solutions of the salts of the metals among the typical elements: think of calcium chloride or zinc sulphate. Further, adding other molecules with lone pairs to aqueous solutions containing transition-metal ions — NH_3, CN^-, OH^-, $C\ell^-$ — often causes more colour changes; something which does not happen with typical-metal salts.

The sequence:

$$CuSO_4 \text{ (white powder)} \xrightarrow{\text{water}} \text{Blue} \xrightarrow{\text{ammonia}} \text{Deep Blue}$$

shows these changes very graphically. Compare with copper's next-door neighbour in the periodic table, zinc, where we have:

$$ZnSO_4 \text{ (white powder)} \xrightarrow{\text{water}} \text{Clear} \xrightarrow{\text{ammonia}} \text{Still Clear.}$$

Clearly there is something happening here: the chances are that what is being seen involves more than simple electrostatic interactions. Surely even these simple observations mean that significant electron redistributions are

---

[5]In fact, even the uncharged nickel atom is so hungry for electrons that metallic nickel will combine with the lone pairs of carbon monoxide (CO) molecules when they pass over the heated metal to form the $Ni(CO)_4$ *molecule*.

occurring. 'Significant electron redistributions' may well be just another name for 'chemical bond formation'.[6]

There is, at present, no coherent qualitative theory of the electronic structure of the bonds in molecules containing transition metals ('complexes'). Existing *quantitative methods* tend to involve the use of symmetry-adapted MOs (see Chapter 8 to describe the electron distribution of the molecule as a whole rather than its identifiable electronic substructures (bonds, lone pairs etc.). On the other hand, qualitative approaches to the subject tend to be more empirical and *ad hoc*, based on chemical experience and using methods familiar from the simpler electronic structures of organic and 'typical element' molecules.

16.4   History and Apology

The theory of the electronic structure of transition-metal-containing molecules presents a real challenge to theorists. It is obvious that some of the familiar concepts of organic and typical-element chemistry still apply, as we saw in discussing the σ bond in the MnO_4^- and OsO_4 molecules in Chapter 13. We can still talk about a 'M–C bond in M–C$\equiv$O' or 'a $3d$ lone pair', for example, but the complex chemistry of many transition-metal molecules clearly must involve more electronic substructures than are found in the simpler chemistry of the typical-element compounds. In order to appreciate the different approaches to the problem of the electronic structure of transition-metal molecules, it is necessary to take a brief look at the history of the subject.

16.4.1   *The 'crystal' model*

The situation is made more confusing by the historical development of the subject of transition-metal chemistry:

* The first quantum-mechanical descriptions of (part of) the electronic structure of these systems consisted not of attempts to explain the *bonding* in the molecules but of attempts to explain the fact that these molecules are often *coloured*. In more scientific terms, the molecules absorb visible light and so have a spectrum in the visible region. But, in

---

[6]There is another question here, 'what actually happens when a metal ion dissolves in water or any solvent?', but this will not be tackled here.

these molecules it is, in the main, the *non-bonding* 'd electrons' which
are involved in the electronic transitions, which are the source of the
colour.

- So attention was concentrated simply on the role of the ligands in influ-
  encing the energy and distribution of the non-bonding electrons, rather
  than the chemically much more important question: 'what is holding the
  molecule together and how is this done?'

The result of this concentration on the properties of the central atom,
rather than the structure of the molecule, was the opinion that an understanding of the bonding was secondary to an understanding of the effect
of the ligand on the energies and distributions of the d-electrons. Tables of
these effects were generated, but little progress was made on understanding
the electronic substructures involved in the constitution of the molecules.
The steps were:

Crystal Field Theory: Concentrates on the electrostatic effects of the ligands on the electrons of the metal atom. Tables of the strengths of these
effects are available.

Ligand Field Theory: Again, as can be seen from the 'field' appearing in the
name, allowance is only made for a qualitative consideration of 'covalent
effects' on the properties of the mainly non-bonding d-electrons. Just as in
the case of Crystal Field Theory the model of the molecule is essentially
ionic: the bonding was *rationalised* as being due to the attraction between
an improbably highly charged central metal atom and charged ligands,
which donate electrons to the central atom not *explained* by a theory of
bonding, which would require electrons to be donated to a *bond*.

16.4.2 *The molecular orbital model*

When it became possible to actually compute the electron distribution
of these molecules there was really only the option of what one might
call the 'symmetry-adapted brute-force' method. Since there was only a
rudimentary knowledge of the type of electronic substructures involved in
these molecules, one had to fall back on the AO-based LCAOMO method,
i.e. use all the occupied and low-energy unoccupied AOs of each atom in the
molecule and simply let Schrödinger's quantum mechanics find out what
happens when the MOs are expanded in this way. This was, in the early
days of quantum chemistry, a challenging task for the computers, and so
the calculation was simplified by using the (often very high) symmetry of
the molecule; the distribution of *each electron* was constrained to have the

full symmetry of the molecule. This means that there are no 'bond MOs' or 'lone-pair MOs' but that each MO is, typically, composed of contributions from *all* the AOs of all the atoms in the molecule which are equivalent by symmetry. We have seen in Chapter 8 that this does give a good description of the *total* electron density of the molecule and of its energy, but at the expense of a 'chemical' description of the electronic structure. There are no chemically-meaningful electronic substructures to be found in this way.

Here is an example, to simply emphasise some similarities between the symmetries of compounds between the N_2 and CO molecules and transition-metal atoms, which uses the localised MOs to illustrate these similarities clearly. The chromium atom forms a number of molecules with CO and N_2; in particular the molecule *trans* $Cr(CO)_4(N_2)_2$ is known as well as the more familiar $Cr(CO)_6$, which is a typical metal carbonyl. In both cases:

- The Cr–C–O and Cr–N–N fragment is linear.
- The diatomic molecule is thought to bond to the metal atom by means of a dative bond from a lone pair — approximately an *sp* hybrid — on the carbon or nitrogen atom.
- The form of the resulting localised MO is a mixture of the HAO on the C or N atom and a HAO on the Cr atom.
- The HAO on the Cr atom is a mixture of the 4*s*, 4*p* and 3*d* AOs: approximately *spd²*.
- The MO contains a much larger proportion of the donating lone-pair HAO than of the metal-atom HAO; a ratio of about 90% to 10%.

The two diagrams below help to appreciate something of the bonding in this molecule. First, contour diagrams of the HAOs involved in the N–Cr dative bond:

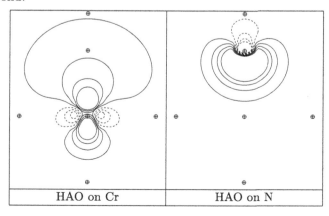

| HAO on Cr | HAO on N |

The HAOs involved in the bond between the C atom of a CO molecule and the Cr atom are very similar indeed, and can scarcely be distinguished 'by eye' from the ones shown for the Cr and N atom above and so are not shown here. They are, of course, pointing along the four Cr–C axes perpendicular to the Cr–N axes.

The diagram below shows both the Cr–N and the Cr–C MOs with contours drawn on a 'slice' through a central plane of the octahedron with the metal atom at the centre. As in the above HAO diagrams, the N_2 molecule is vertical and the CO molecule is horizontal. The diagrams below of the localised MOs are also very similar. They are linear combinations of the N, Cr and the C, Cr HAOs respectively and are clearly 'mostly' the N or C HAO as we noted above. The C–Cr bond MO extends very slightly over the Cr atom, which perhaps gives a hint that it is the stronger bond of the two. Of course, there *are* differences between an N–Cr bond and a

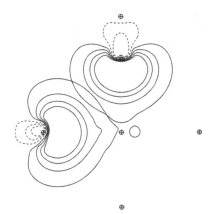

C–Cr bond not least the very much smaller bond energy of the $Cr–(N_2)$ bond than the $Cr–(CO)$ case. Naturally, this is not visible on the contour diagram.

However, there are three main points to be made from this single example:

- The fact that the use of symmetry orbitals completely masks the actual bonding scheme in the molecule.
- We see our main idea — the existence of environment-insensitive sub-structures — in the electronic structure of at least these molecules very clearly.

• There is only a small contribution to the bond MOs from the HAOs of the chromium atom 'pulling' the electron distribution away from the ligand towards the metal atom. The very unequal contributions of HAOs from the two centres involved in both bonds is very different from the structure of the σ bonds in organic molecules. If the contributions were equal within the bonds this would mean a very large electron population on the central chromium atom due to the six similar bonds. In fact, if the electron populations of the atoms in this molecule are calculated by the method described in Section 14.3, the chromium atom in the molecule has an electron count of about 24.2, that is, only 0.2 electrons over its neutral-atom count.

This is just one molecule which perhaps contains six similar bonds and may well not be at all typical of the electronic substructures in transition-metal molecules.

16.4.3   The 'chemical' model

In the 1960s and after, there was an explosion in the subject of transition-metal chemistry with the discovery of a vast range of organometallic compounds: compounds with large and varied *organic* molecules as ligands, in place of the atomic and simple molecular ligands. In these cases there was no question of treating the organic-to-metal binding as ionic. The natural assumption was that, in many cases, there was a familiar two-electron covalent bond albeit perhaps datively formed. More exotic species in which conjugated organic molecules bonded 'face on'[7] to a metal were found, again showing that the ionic model was totally inadequate to describe the nature of transition-metal bonding.

In these circumstances practising chemists will simply *demand* a model of the electronic substructures which takes account of their findings and, if one is not available from the theorists, will create one of their own. If, therefore, there is no evidence that the bond between chromium and carbon in $Cr(CO)_6$ involves delocalised electrons over all six Cr–CO fragments, but all the experiments point to a two-electron bond as a good starting point then the model taken from typical-element or organic chemistry will be used. Why not treat this bond from CO to another centre in the same way

---

[7]Ferrocene and other metallocenes are the most spectacular examples.

as we did in Chapter 10: a datively-formed single electron-pair bond:

$$\text{If BH}_3 + \text{CO gives BHCO } (\text{H}_3B \longleftarrow \text{CO})$$

then why should

$$\text{Cr} + \text{CO not give Cr}-\text{CO } (\text{Cr} \longleftarrow \text{CO})?$$

and so on. Further, why should a metal-ligand bond *always* be described as dative? Does anyone *really* think that the MnO_4^- molecule is *really* composed of dative bonds between four O^{2-} ions and an Mn^{7+} ion? Mn^{7+} is the sort of thing that might be found in a particle accelerator or in the atmosphere of a star, not in an innocent-looking purple solution in water!

So, we find among practising chemists the use of HAO-based localised bonding schemes in their descriptions of the bonding and reactions of transition-metal compounds. The unfortunate thing about this situation is that these models tend to make no reference at all to the laws of nature.

16.4.4   *Apology*

The upshot of the situation summarised above is a state of affairs in which the theoretical description of the electronic structure of transition-metal molecules has split into two:

The Symmetry-adapted MO Model[8]: This model requires a host of 'effects' to come close to explaining the *chemistry* of the molecules: 'orbital interactions', 'repulsion of energy levels', for which there can be no experimental evidence, since orbitals cannot interact and energy levels, being mathematical functions, cannot repel each other; only charged particles can interact or repel.

The Empirical Chemical Model: This approach is based simply on ideas which work for typical-element chemistry, which uses concepts based on localised HAOs and conjugated π systems; the sort of structures which have given sterling service in organic chemistry.

And, of course, mixtures of both systems.

---

[8]Note that the use of delocalised *MOs* does *not* imply delocalised electronic substructures, an additional source of confusion: see Chapter 8.

Clearly what is required is a study of the structure and reactions of transition-metal molecules with a view to identifying the types of electronic substructures which can help explain the enormous range of properties of these molecules. The structures we have so far identified — electron-pair bonds, lone pairs, conjugated π systems, atomic cores — are still valid and do occur in these molecules, but what is clear to anyone at all familiar with transition-metal chemistry is that these structures are not sufficient to give a full account of the range of molecular structures and behaviour. This task is one of research, not teaching, and this is certainly not the place to even begin to carry it out; it will not be attempted here. So an apology is in order, for perhaps promising what cannot yet be given. The development of a qualitative theory of the electronic substructures to be seen in transition-metal-containing molecules is a rich area for study, demanding a detailed knowledge of quantum theory and the chemical structure and behaviour of these important systems.

16.5 Comments

Anyone studying chemistry is likely to find that there are two main approaches to the qualitative quantum theory of the electronic structure of molecules. Roughly speaking these are:

(1) The symmetry-adapted LCAO method, which is used mainly by inorganic chemists for the historical reasons summarised at the start of this chapter, and by computationally-inclined chemists, because it can be computer-implemented very easily and involves no input describing the substructures in the molecule.[9]

(2) The empirical extension of the HAO-orientated method used in typical-element chemistry, which is loosely based on the valence bond (VB) model described briefly in Section 5.5.

The connection between these two approaches is not easy to see and involves a far more detailed knowledge of formal quantum theory than is possible to give here. This split is very unfortunate because these two approaches are taught independently, with little emphasis on the fact that they are both trying to explain the same phenomena. The drawbacks of

---

[9]What I have, rather disparagingly, described as the 'brute force LCAO method'.

each method are extremely obvious:

- The LCAO method requires no input describing the electronic substructures involved in the molecule (bonds, lone pairs etc.) and therefore generates no output describing these structures.[10] But, by putting this information into the *results, after the calculation has been performed,* one can extract some quantitative information about the details of the substructures. The result of a typical LCAO calculation is, usually, a set of MOs which are spread out over the whole molecule; they may have contributions from all the atoms in the molecule. This has the unfortunate effect of seeming to imply that the electronic substructures are themselves delocalised like the conjugated π systems. The problem is they may be or they may not be, as was emphasised in Section 11.2 when contrasting the σ and π structures of benzene.
- The empirical method is just that: empirical. It makes no reference to the known laws of nature and so is completely adrift and can use whatever additional explanations and whatever 'effects' that seem to be needed to 'rationalise' the observed facts.[11]

> It is up to chemists to decide — on the basis of chemical experience — what the likely electronic substructures are within a molecule; simply solving the equations for the distribution of electrons cannot do this, as Chapter 1 stressed. Nor can a system of rationalisation which makes no reference to the laws governing the distribution of electrons provide anything but a temporary stop-gap.

What is very likely to be happening in the electronic structure of transition-metal compounds is that the basic model which has been used throughout is perhaps starting to break down: these electronic substructures are not so 'environment insensitive' as they have been assumed to be so far. This should not surprise us. The model is just that: a model. One of the most interesting (and useful) properties of transition-metal molecules is their catalytic capabilities. They seem to be able to persuade molecules to undergo chemical changes which they stubbornly resist if left

---

[10] Remember that computers are not intelligent, they can only do simple, routine things very fast!

[11] It is, perhaps, significant that in this context one 'rationalises' rather than 'explains'. Nature already *is* rational, if we think it needs 'rationalising' it is our understanding which is in error, not nature.

to their own devices. What must be happening here? Remember, chemical reaction is just the rearrangement of electronic substructures — breaking of some bonds and the formation of others — and this, in itself, is a break-down of the idea of stable electronic substructures; bonds are broken and bonds are formed.

The traditional empirical definition of a catalyst is something like:

> A catalyst is a substance which speeds up a chemical reaction while not changing itself.

This presents a very idealised picture, as if a catalyst were some kind of magic elixir whose mere presence caused a chemical reaction to occur. The catalyst *must* be involved in the chemistry of the process, and, what-ever the detailed mechanism, what must happen is the formation of some intermediate between reactants and catalyst-containing bonds which are easily transformed or broken after some internal transformation in the intermediate, thus restoring the catalyst. If the independence of the elec-tronic substructures in transition-metal molecules is much less than in (e.g.) organic molecules then, once an intermediate molecule of this type is formed, transformations become easier.[12] That this is so is nowhere more obvious than the catalytic action of iridium salts in the reaction mentioned on page 183:

$$CH_3OH + CO \longrightarrow CH_3CO_2H.$$

The catalyst is put into the industrial reaction vessel as nice clean crys-tals, and comes out months later as a brown slurry to be sent off for re-processing into crystals again. The iridium is still there all right, but is hardly 'unchanged'.

The conclusion is that, in the main, we cannot give any conclusions because the electronic structures of transition-metal-containing molecules are so varied and so complex that only the most elementary ideas developed here can be applied.

16.6 Assignment for Chapter 16

One of the most easily prepared and famous molecules involving the ethene entity acting as a 'donor' to form a dative bond is Zeise's salt, or rather

---

[12]But still a complex problem for the chemist to elucidate!

the anion involved in Zeise's salt. This anion was discovered in 1825, is pale yellow in colour, contains platinum and has the empirical formula $[Pt(C_2H_4)C\ell_3]^-$. The ion was a conundrum for almost a century since the theoretical tools for interpreting its structure were not developed until the electron distribution in a double bond was discovered. It is now known to be a (roughly) square-planar compound with the platinum atom in the middle. The corners of the square are occupied by the three chlorine atoms and an ethene molecule. The ethene molecule is attached to the central platinum 'face on'; one lobe of the π MO between the two carbon atoms is thought induce the π-MO to act as a donor to form a dative bond in a similar way to the action of the major lobe of a lone pair:

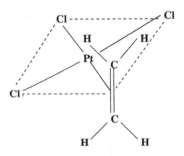

The CC distance is very close to the corresponding CC length in ethene itself.

If, instead of ethene, tetracyano-ethene is used to form an analogous molecule we find that the distance between the two C atoms is larger and is close to the C–C distance in ethane. In this molecule the CN groups are significantly bent away from the Pt atom.

What is happening here? Think about the 'non-bonding' interactions between the CN groups and the lone pairs on the Pt atom.

Chapter 17

Omissions and Conclusions

The electronic structure of matter is a very complex subject, and what has been attempted in this work is basically an approach in which the physical interpretation of the results of the application of Schrödinger's mechanics (quantum theory) is at the forefront. There is much work still to be done, particularly in the inorganic chemistry area where, as yet, only the barest outline is possible. This chapter describes some areas which have been deliberately omitted so far, and tries to draw together the ideas which have proved fruitful in the study.

Contents

There are a number of important areas in the theory of the electronic structure of molecules which have been omitted from this work, mainly because they involve the *interactions* between molecules, which is a large and developing area and which involves more advanced techniques of quantum theory.

17.1   Omissions

An interpretation of the electronic changes involved in the interactions between molecules — in particular, chemical reactions — is a central theme of chemistry, which presupposes a robust qualitative and quantitative theory of the electronic structure of the individual molecules which interact or react. With the exception of a brief look at the hydrogen bond, this area has not been examined.

17.1.1   *Intermolecular forces*

The central problem with the theory of interactions between molecules which do not react is that the forces and energies involved are much smaller than those involved in chemical bonds, and very much smaller than the total energies of the molecules themselves. In Section 6.3 we saw that the repulsions between adjacent electron pairs are huge: hundreds of kilojoules per mole. Interactions between molecules A and B, say, involve all the attractions between the nuclei of A with the electrons of B, the repulsions between the nuclei of A and B, and the repulsions between the electrons of A and B. Now, while these individual components of the energy of interaction between the two molecules may be large, they almost completely cancel the other out, leaving the actual interaction energy as a small difference. The problem is, if one has the energy of A and B and the energy of AB, the difference:

$$\Delta E = E_{AB} - (E_A + E_B)$$

is very sensitive to errors in E_{AB}, E_A and E_B. It has been compared to weighing the captain of a ship by weighing the ship with him aboard and again when he has disembarked and taking the difference.

   In this situation, one has to use much more refined methods of computational quantum chemistry to obtain even a sensible answer, and the accurate calculation of these numbers is very challenging. While one can make a guess at the types of phenomena involved in intermolecular forces it is not at all easy to provide a qualitative picture of these interactions.

17.1.2   *Chemical reactions*

In the case of reactions between molecules we have, at least, some knowledge of the electronic structures of the reactants and products. That is, we know what the electronic substructures are before and after the reaction. We have already seen in Section 6.4 that reactions leading to a datively-formed covalent body are particularly easy to understand, since there is no bond broken and the two reacting species are only too eager to come together by simple electrostatic attraction. The problem, then, is to try to see *how*, in the more interesting type of reaction where one bond is broken and another one is formed, the substructures are deformed and changed.

In general, there is a *barrier* to chemical reaction, in the sense that the two reactants can co-exist without reaction until they are given some kind of input of energy (heat, say) to induce the molecules of the reactants to collide with enough force to overcome this barrier. When this is done, there is the re-arrangement of electronic substructures in the vital parts of the reactants to form the new bond and hence the products. At first sight, the bonds which are most vulnerable to being broken are the ones for which the electrons are least tightly-bound to the atomic cores; the ones in the MOs with the highest energy levels. This will not always be the case, since we have already seen that the electrons involved in forming a dative bond are often lone pairs, which may or may not be the least tightly bound. Certainly in the case of the CO molecule the lone pair is more tightly bound than the π MOs but, of course, it is 'pointing' in a direction which makes it very attractive to other systems with areas of predominantly positive charge.

The idea that the highest occupied MO (HOMO) of one of the reactants will be the one which will most readily have its electrons re-arranged or even removed has been made the basis of a theory of the reactions of aromatic compounds. Partly because the theory was developed before it was possible to perform accurate calculations on the σ systems and, more importantly, because these systems are the most suitable for such treatment:

- The simplest type of MO calculation is one in which:

  (1) The electronic structure really is delocalised, so that the symmetry-adapted MO method generates chemically realistic MOs for the system.
  (2) The most difficult numerical problem — the calculation of the effects of electron repulsion — is omitted altogether from consideration.

This is most suitable for the delocalised π systems of aromatic and conjugated molecules.

- If electron repulsion is omitted from the calculation, the MOs of the cations and anions are the same as those of the parent molecule. That is, the addition to and removal of electrons *does not change the MOs*.

This elementary model, involving only the interactions between electrons in two MOs — the highest occupied MOs of A and the anion[1] B^{-2} — has been widely used to rationalise reactions involving molecules with delocalised π substructures.

When we consider the case of reactions involving localised (σ) structures the situation is very different:

- Electron repulsion is important, even decisive, in these cases.
- Typically, only a small part of the molecule is involved in the chemical change, so that localised substructures are involved.
- It is not always the least tightly-bound electrons which are crucial in the reaction.

Perhaps the simplest illustration is the archetypical S_N2 reaction of the hydroxyl radical with methyl iodide:

$$HO^- + CH_3I \longrightarrow CH_3OH + I^-.$$

The highest occupied MOs of the OH^-, by far, are the degenerate pair of π MOs, which are the O atom $2p_\pi$ AOs basically unchanged by molecule formation. The MO which contains the electrons that will be involved in the reaction, forming the initial attack on CH_3I to form the C–I bond, is, of course, the σ (sp-like) lone-pair MO pointing away from the H–O σ bond.

This situation is quite common. It is not necessarily the least tightly-bound electrons which are in the front line of the chemical attack, but those which are in the right place, in the right localised substructure and, presumably, the ones whose distribution is most easily polarised on approaching the reacting partner. This almost always excludes the symmetry-adapted delocalised HOMO computed by standard techniques. What seems most likely to happen during a simple one-step chemical reaction is something

---

[1]The highest occupied MO of B^{-2} is often called the lowest unoccupied MO of B, and is said to interact with the HOMO of A. As we have noted before, orbitals (occupied or 'unoccupied') are mathematical functions and cannot interact.

similar to the schematic steps described in the formation of a chemical bond from atoms:

- The approaching systems polarise each other's electron distributions.
- New transient electron substructures are formed, possibly involving partial bonding between both systems (a 'transition state').
- The reorganisation of the electron distribution into new localised structures involving the new bond.

Work along these lines has been done, particularly by Japanese workers, and involves significant computational analysis.

17.2 Conclusions

It is sometimes jokingly said that the conclusions to a piece of work should be the same as the objectives set out in the introduction but in the past tense. We cannot be quite so smug here since, throughout the work, there have been new types of electronic structure 'turning up'.

The main conclusion is, however, a reinforcement of the original idea that the study of any complex system in nature is only possible if one can at least start from a breakdown of the system into smaller sub-systems which are as independent from each other as possible. We have to make an *idealised* structure from the real, complex mixture.[2] What these substructures actually are cannot be answered by thinking very hard about the problem; it can only be found out by *observation and experiment*, by studying the behaviour of real molecules. Just when it looks as though we are beginning to get to grips with the possible electronic substructures of molecules, something completely new and inexplicable turns up: boron hydrides, ferrocene, bridging structures, clusters etc. That it is, in fact, possible to see environment-insensitive substructures in molecules which consist of many charged particles in mutual interaction with strong long-range forces between all of them, is, if one could look at it without prior knowledge, something of a miracle. The analogy in Chapter 1 with the sub-systems of the solar system — the individual planets and their moons — is useful but the forces of gravity are billions of times weaker than electrostatic interactions. What is important is to always bear in mind that this, or any other,

---

[2]Not so bad, perhaps, as the extremely simple abstractions which economists use to 'model' a real economy!

model is just that, a model; it is not nature, it is in our heads to help us to cope with complexity; it will, sooner or later, break down and prove inadequate. Einstein, whose 'God' was Nature, said, 'The Lord is subtle but not malicious'; we have to hope that is true.

What we have tried to do is concentrate on this model of molecular electronic structure: the way in which the substructures which are more strongly interacting within themselves than in their interactions with others. With the simple exception of the VSEPR method in Section 6.3, the effects of the interactions amongst bonds and lone pairs have not been given any detailed attention; this becomes vital in any theory of chemical reactions and dynamics.

It has also been necessary to try to combat the tradition in chemistry of relying on many rules-of-thumb which, in time, tend to become substitutes for physical interpretation and explanation. Chemists are proud to be at the 'centre' of physical science, with one foot in physics and one in biology, but, for this to be most useful, the theories of chemistry must be intelligible to others and not a private system of rules. What is surely true is that, once we have made a model, then the only help we have with using and investigating that model is the combination of experiment, Coulomb's law, Schrödinger's mechanics and Pauli's principle.

17.3　Assignment for Chapter 17

Write a brief evaluation of your experience in using this book and send it to me (d.cook@sheffield.ac.uk). You can be as critical (or as complimentary!) as you wish. Suggestions for improvement would be very welcome — I am particularly keen to know if you thought the assignments were useful.

Appendix H

The Pauli Principle, 'Spin' and Electron Repulsion

Contents

H.1   Pauli and Electron Interactions

There are some important consequences of the Pauli principle which occur throughout the theory of atomic and electronic structure. This short appendix tries to explain how the effects of electron spin and electron repulsion have been confused since the very early days of quantum theory. We have given the simplest possible 'working' statement of the principle but there are more abstract, general statements from which our definition can be derived as a special case. We can look at the physical interpretation of the principle by thinking about probabilities and densities.

The probability distribution associated with an electron in an orbital ($\phi(\vec{r})$, say) is just the function of space given by the square of that orbital:

$$\rho_1(\vec{r}) = [\phi(\vec{r})]^2, \tag{H.1}$$

where the symbol $\vec{r}$ has been used to represent the co-ordinates of the electron occupying the orbital. When we consider the repulsion between electrons we should think about the *relative* distribution of electrons. So, for example, we need to know the distribution of two electrons; where electron 2 is when electron 1 is at a particular point in space. In other words, we need a function ($\rho_2(\vec{r}_1, \vec{r}_2)$, say) which gives the probability that if electron 1 is at point $\vec{r}_1$, then electron 2 is at $\vec{r}_2$. But, from the orbital description of an atom or molecule, we will usually have some function composed of AOs of the two electrons, so we may write

$$\rho_2(\vec{r}_1, \vec{r}_2) = [\Psi(\vec{r}_1, \vec{r}_2)]^2, \tag{H.2}$$

where the two-electron function $\Psi(\vec{r}_1, \vec{r}_2)$ is some product of AOs.

Since electrons are all the same — they have been labelled as 'electron 1' and 'electron 2' for our convenience — they do not know we have labelled them! This means that if we interchange them then the probability must be unchanged; the probability that electron 1 will be at one fixed point and electron 2 will be at another fixed point cannot change if we swap the two identical particles. So:

$$\rho_2(\vec{r}_1, \vec{r}_2) = \rho_2(\vec{r}_2, \vec{r}_1). \tag{H.3}$$

An interesting point arises when we consider what happens when two electrons 'collide', i.e. they are both in the same place at the same time. This happens, of course, when $\vec{r}_1$ has the same value as $\vec{r}_2$. If we put

$$\vec{r}_1 = \vec{r}_2 = \vec{a} \ (say), \tag{H.4}$$

where $\vec{a}$ is the co-ordinates of a single fixed point, it is obvious that:

$$\rho_2(\vec{a}, \vec{a}) = \rho_2(\vec{a}, \vec{a}), \tag{H.5}$$

which is just a special case of the above equation.

But, when we ask for a similar answer to the question 'what happens to the two-electron function $\Psi(\vec{r}_1, \vec{r}_2)$' the answer is ambiguous because both Ψ and $-\Psi$ given the same function ρ_2 when squared, so that:

$$\Psi(\vec{r}_1, \vec{r}_2) = \pm\Psi(\vec{r}_2, \vec{r}_1). \tag{H.6}$$

In particular, for two electrons colliding:

$$\Psi(\vec{a}, \vec{a}) = \pm\Psi(\vec{a}, \vec{a}), \tag{H.7}$$

which looks harmless enough. But there are *two* solutions here, the one that one might expect — with the plus sign — but also one with the minus sign.

Now if

$$\Psi(\vec{a}, \vec{a}) = -\Psi(\vec{a}, \vec{a}) \tag{H.8}$$

(special case of the simple algebraic equation $x = -x$), then there is only one solution:

$$\Psi(\vec{a}, \vec{a}) = 0, \tag{H.9}$$

that is:

> If the two-electron function changes sign when two (identical) electrons are interchanged then the electrons can never collide.

In fact, the more general statement of the Pauli principle says that this is precisely what is the case:

> Any Ψ function which describes the distribution of two or more electrons must change sign when the co-ordinates of any pair of electrons are interchanged. Or, more compactly, 'the Ψ function for a many-electron system must be *anti-symmetric* with respect to exchange of electrons.'

Now, if the Pauli principle alone causes the electrons to avoid each other, this must have an effect on the size of the energy of repulsion among a set of electrons in an atom or molecule.[1] A constraint which gives a reason for electrons to avoid each other *in addition to* their mutual repulsion will obviously *reduce* the electron-repulsion energy compared to that for which no such constraint was present.

This is where the idea of electron 'spin' complicates the issue.

H.2 Effects of Electron 'Spin'

So far, the Pauli principle has not given any more basic explanation of *why* a maximum of only two electrons may occupy a given orbital. This question involves the idea of giving each electron in a many-electron system its own unique characterisation. It has been mentioned several times that electrons have, in addition to their electric charge, a *magnetic moment* — they

---

[1] Of course, the fact that the electrons both have the same electric charge causes them to avoid each other as well.

behave like very small magnets which can take up only one of two possible orientations in space. On an everyday scale an electrically-charged body which spins generates a magnetic field, so, by analogy, charged electrons which have a magnet moment are said to be spinning in one of two possible directions.[2]

So, in order to completely specify a given electron:

- In space, we must give three spatial co-ordinates (x, y, z, for example) *and* the direction of its spin.
- In an orbital description, we specify the orbital it occupies *and* the direction of its spin.

We can now give a more complete statement of the Pauli principle by adding a few more words (here, in italics) to the definition in Section 2.1.3.1:

> When a many-electron system is described by electrons occupying orbitals, no more than two electrons may occupy any one orbital. *If an orbital is doubly-occupied, the two electrons must have opposite spin.*

The introduction of the idea of specifying an electron by its spin as well as its spatial co-ordinates is where the Pauli principle brings in a connection between spin and electron repulsion.

In the equations of the previous section the symbol $\vec{r}$ was used instead of explicitly giving a set of spatial co-ordinates (x, y, z) in order to make the transition to include electron spin in the specification of an electron easy. We can now include the value of an electron's spin and simply say that the symbols $\vec{r}_1$ and $\vec{r}_2$ are generalised and now mean, respectively, the spatial co-ordinates *and* the spin direction of each electron:

$$\vec{r}_1 = (x_1, y_1, z_1; s_1) \qquad \text{(H.10)}$$

$$\vec{r}_2 = (x_2, y_2, z_2; s_2) \qquad \text{(H.11)}$$

$$\vec{a} = (a_x, a_y, a_z; s) \qquad \text{(H.12)}$$

and the argument still applies but with a new twist — equation (H.8) must include the fact that the electrons must have the same *spin*. So, only electrons *with the same spin* are constrained by the Pauli principle to never

---

[2]We have seen that this is just an analogy in Appendix E.

collide; electrons with opposite spin are not constrained at all by the Pauli principle in its newer form.

This means that the electron repulsion between two electrons with the same spatial distribution (e.g. in the same orbitals) will repel each other differently depending on whether they have the same or different *spins*. In fact, the repulsion between electrons of the same spin is *always less than* that between electrons of different spin.[3] This is certainly a startling result; an actual energy of interaction can depend on something which is not included in the energy expression.

This result gives an immediate explanation of Hund's rule mentioned in Section 2.4. The example was the electron configuration of the carbon atom in which the two electrons placed in separate AOs due to electron repulsion have the same spin rather than opposite spins, since this smaller effect is due to the Pauli principle, which we have just discussed. More generally, it gives an explanation of why electrons in degenerate orbitals tend to have the same spin in an apparent contradiction to the way they would line up if the interaction between them were due solely to their 'genuine' interaction via their magnetic moments.

H.3   Three Types of Spin Alignment

This is perhaps a good place to distinguish the three possible cases of the mutual alignments of electron spin:

(1) Considered simply as *magnets*, it is obvious that electron spins will take up the same alignment to each other as any other pair of magnets: north of one magnet as close as possible to south of the other, and south of the first magnet close to the north of the other. If there are only two possible ways in which the electron spins can be orientated, clearly this will be where the two magnets are *antiparallel*.

(2) If two electrons occupy different *degenerate* orbitals (that is, the spin alignment is the only way the energy of the two can be changed) then, as we have seen above, the Pauli principle ensures that the two spins will be *parallel*.

---

[3]In orbital language, this means that an additional energy term — the exchange term — is subtracted from the ordinary repulsion term and, since this exchange term is always positive, the repulsion between electrons of the same spin is always less than the repulsion between electrons of different spins.

(3) Finally, if the two electrons of a chemical bond are responsible for a contribution to holding two atomic cores together then, of course, they will both need to be in the *same* lowest-energy orbital available. If they are to be in the same orbital, the Pauli principle again is responsible for their mutual spin alignment. Any two electrons occupying the same orbital must have spins which are *antiparallel*.

Only the first of these cases is the actual spin-magnetic interaction between the electrons involved. This magnetic interaction is tiny compared to the strong electrostatic attractions and repulsions among the nuclei and electrons, and plays no part at all in the energies of chemical bonds.

The third case is of interest because it is sometimes said that the *driving force* for chemical bond formation is the *antiparallel pairing* of electron spins. But this is to put the cart before the horse. As we have just seen,[4] the fact that the spins of the electrons involved in a chemical bond are antiparallel is an *effect* of bond formation, not its cause. A two-electron chemical bond would have the same energy whatever the orientation of the spins of the two electrons (if the Pauli principle allowed it!), because it is the electrostatic interactions which lead to bond formation.[5]

---

[4]This point is elaborated in Appendix E.

[5]Of course, we could not, in fact, turn off the Pauli principle for just two of the electrons in a molecule!

Appendix I

A Note About 'Unoccupied' MOs

Almost nothing has been said about the rôle of unoccupied MOs. Since some theories of the reactions between molecules make use of these entities, it is worth while looking at how these MOs occur and their physical interpretation.

Any calculation of the MOs of a molecule having $2n$ electrons will usually use more than n HAOs in order to construct the n MOs which are needed to accommodate the $2n$ electrons.[1] In fact, we have already seen earlier that, if any combination of n orbitals is used to form a new set of the same number (n) of orbitals, this does not change the electron distribution or the total energy of the system.[2] So, if we use m HAOs to form n MOs, (where, of course, $m > n$) we can generate m MOs, which leaves $(m - n)$ 'unused'. These MOs are called, for obvious reasons, 'unoccupied MOs'; a better term, as we shall shortly see, is 'virtual orbitals'.

When we look at the energies and detail forms of these orbitals, two things emerge:

(1) In contrast to the MOs which are occupied (used to describe the electronic structure of the molecule), they have *positive* energies.
(2) They are usually much 'lumpier' than the occupied MOs — they typically have little electron density between nuclei which are bonded.

---

[1]Two electrons per MO, remember.

[2]Recall the two different pairs of lone-pair MOs in methanal, and the two ways of describing the cores of any molecule.

An MO which has a positive energy means that, if electrons were to occupy it, that electron would not be bound to the molecule — a molecule in this condition would spontaneously eject those electrons, i.e. it would spontaneously ionise! This is a rather surprising result and, at first sight, seems to be contradicted by the fact that the negative ions of some atoms and molecules do, in fact, exist; for example both Br and Br$^-$ do occur.

There are several factors involved in this apparent disagreement between experimental facts and the MO theory:

(1) In some, but not all, cases, if a *full calculation* is carried out on the anion, i.e. using the actual MOs of the anion and not those of the parent atom or molecule, all the occupied orbitals have negative energies. This is obviously because the presence of the extra electron changes the distributions of the other electrons and stabilises the anion; another case of the importance of including all the electron repulsions.

(2) The point has not been stressed earlier but, when we carry out a MO calculation, we do so on *the isolated molecule* — one molecule on its own in outer space! The same applies to any MO calculation on an anion. Now, our experimental observations, particularly of ions, are not very often carried out on atoms and molecules in outer space or even in the gas phase. Most observations involving ions are those in which the ions are in some *polar solvent*, often water. The solvent molecules stabilise the ion by providing them with an environment of the opposite charge.

(3) The most general reason, however, lies in the way we choose to do the calculation. The use of HAOs as building blocks for MOs implicitly assumes that the electron density in the molecule is not too different from that of the separate atoms, i.e. concentrated close to the atoms and between the nuclei. The additional electron(s) in anions experience a much weaker net attraction to the nuclei than those in the neutral molecule because the total charge 'seen' by these electrons is very small: the sum of the nuclear charges and the electrons in the neutral molecule is *zero*. Therefore we are not routinely providing the right 'kind' of orbital to be used to describe these loosely-bond electrons.

If we take special steps to rectify this deficiency in the HAOs provided, i.e. include very diffuse orbitals in the calculation, it is often found that the lowest unoccupied 'MO' is an orbital whose energy is close to zero. What this is telling us is that, for an isolated, unstabilised anion the preferred, lowest-energy state is the neutral atom plus a free electron; an orbital energy of zero means that the electron is free.

There is every reason, therefore, to avoid giving much importance to the energies and spatial forms of the unoccupied MOs which are generated as a by-product of the MO calculation; sometimes they may be meaningful, sometimes they are not.

It is worth while to point out that, if a calculation of the electronic structure is carried out *numerically*, that is, solving the equations not by using (H)AOs as building blocks but carrying out the solution by using just a set of numerical values of each MO at points in space — a process which is extremely expensive for molecules — then *no unoccupied orbitals are generated*.

Help with the Assignments

Here are some thoughts about the end-of-chapter assignments which should help with any discussions. The material here becomes less and less detailed as the work progresses since, after a while, the assignments are purely investigative or for discussion and consolidation of the material in the text of each chapter. Consequently, there is no help here for the assignments of Chapters 8, 10, 11, 12, 13 and 15.

Chapter 1

(1) Since Newton, the explanation of Galileo's result is easy to demonstrate. Newton's familiar second law

$$F = ma$$

relates the acceleration, a, to the Force, F, applied to a body of mass m. In the particular case of gravity, the downward force is given by

$$F = mg$$

where g is the acceleration due to the earth's gravitational pull. So that

$$mg = ma$$

and the m cancels, so all masses suffer the same acceleration and therefore take the same time to fall to earth.

As regards the question, in the cathedral square, your observations and experiments might include:

- Noticing that the square in which the cathedral and tower stand has pigeons flying around, shedding feathers. Both pigeons and feathers are massive bodies, why do they not fall?
- These days, tourists tend to drop litter, which sometimes get blown up into the air — paper has mass, how is it able to defy gravity?

Typical experiments to counter Galileo might be to drop:

- A small lead weight and a feather of the same mass.
- Two pieces of aluminium, one a small solid piece and the other of oven foil.
- A small stone and a helium balloon of the same mass.

All of these 'exceptions' to the law depend on the use of the atmosphere to help defy gravity; some actively — birds flying — and some more passively — air resistance. Galileo used *dense bodies* for which air resistance is negligible. So a more precise formulation of the law would be:

> All bodies fall to earth in the same time *in a vacuum*.

What is the moral of this story?

Of course, the moral is:

> No scientific laws are *universal*, they all are limited in some way. The *conditions* under which the law is valid are every bit as important as the law itself.

Chemistry is a complex subject and we shall often have to bear this important fact in mind.

(2) Both these views of the structure of molecules are correct, each for its own *area of applicability* — do not allow disagreements between chemists doing different things to cause useless arguments.

Do not be seduced by the simplest arguments; remember, 'If things were as they seem to be, science would not be necessary':

The earth is not flat.

The sun does not go round the earth but the moon does, and in the opposite direction to appearances.

Water is not a continuous fluid.

Light does not always travel in straight lines, etc.

Chapter 2

(1) First of all think about the simplest case: the lithium atom and the low-est two possible electron configurations, $1s^2 2s$ and $1s^2 2p$. The energy of the $2s$ and $2p$ AOs are determined by the attraction to the nucleus and the repulsion of the two $1s$ electrons. Since the distribution of the $2s$ and $2p$ electrons in space are *different* it is extremely probable that the factors in their energies will be different. The $2s$ electron distribution overlaps with the $1s^2$ distribution more strongly than the $2p$ electron does and is, on average, closer to the nucleus, so its energy is lowered with respect to the $2p$. Check that you agree with this assessment.

The other comparisons follow in a similar way; in summary:

(a) The hydrogen atom has only one electron, all other atoms have several electrons.

(b) These electrons repel one another and the repulsions between the electrons depend on the size and shape of their distributions in space, i.e. on the size and shape of the AOs.

(c) For example, the main difference between a $3s$, $3p$ and $3d$ AO is that the number of lobes in the AO increases in the sequence $3s < 3p < 3d$. Among other things this means that *for a given n* (3 in this case) the $3s$ has a greater density close to the atomic core than the $3p$, which in turn has a greater density than the $3d$. This naturally means that in the sequence $3d\ 3p\ 3s$, an electron in these AOs experiences more and more attraction to the atomic core, and so that is the sequence of their total energies.

(2) This is just the first introduction to the puzzling and apparently con-tradictory behaviour of electron spin in quantum chemistry. Take a look at Appendix H and discuss the problem with your tutor.

(3) This is a further example of trying to give all the symbols in a chemical structure diagram a meaning.

Chapter 3

(1) It is absolutely crucial that the symbols used to represent quantities are understood and become familiar. At the level we are using here mathematical symbols are just shorthand. The expression on page 62 is initially written more or less in words but still contains the symbols

2*s* and 2*p*, the first of which stands for 'the second-lowest spherically-symmetrical atomic orbital, which is a function of ordinary 3D space described by the mutually-perpendicular co-ordinates (x, y, z)' with a similarly long-winded description of the 2*p* AO. No-one can cope with this kind of verbose description and manipulate the expressions in which they would occur. You are familiar with all kinds of short-hand in your everyday life, with which usage has made you familiar; the button on your audio device which allows the sound to be 'paused' just has a symbol on it, it does not say 'Press this button to Pause and Restart play'. So it is with mathematical symbols; for us, they are simply short-hand.

It is worth while giving a warning here about the rather 'non-standard' way which functional relationships are denoted throughout quantitative chemistry. In mathematics one indicates functional dependence with an expression like

$$y = f(x),$$

meaning that there is a *unique rule* from which, given a value of x, one can calculate the corresponding value of y; for each value of x, $f(\ )$ gives a value for y. Often in chemistry we see expressions like $V(p)$ for the volume of a perfect gas at constant temperature as a a a function of pressure, and $V(T)$ for the volume as a function of pressure at constant temperature. The equation of state of an ideal gas is

$$pV = nRT$$

in the usual notation, so we have

$$V = \frac{nRT}{p}.$$

At constant pressure (p_0, say) this gives

$$V = \left(\frac{nR}{p_0}\right) \times T = f(T), \quad (say)$$

while at constant temperature (T_0, say) we have

$$V = (nRT_0) \times \frac{1}{p} = g(p), \quad (say).$$

Now, if we use the same symbol (V) to represent the functional dependence of volume on *both* of these two quantities, strictly we are misusing the notation because $f(\ )$ and $g(\ )$ are completely different functions. So, when we write $V(T)$ or $V(p)$ in this context all it really means is that V (the volume) depends only on T (or p) in the particular circumstances, and we have to be careful with interpreting the notation as mathematical expressions.

In the context of quantum chemistry, one often sees a similar usage in indicating the co-ordinate system in which AOs are expressed — a $2p$ AO might be written as either $2p(x, y, z)$ or $2p(r, \theta, \phi)$ depending on whether one is using ordinary Cartesian co-ordinates or polar co-ordinates.

All these usages are convenient short-hand rather than mathematical notation.

(2) This is revision material — refer to Section 3.2.2.

(3) Since the central assumption behind the use of HAOs in describing the electronic structure of molecules is 'the electron distribution in molecules is not *very* different from that of the separate atoms', any systems for which this is suspect would be candidates here; electrically-conducting metals are the most obvious case.

Chapter 4

(1) There are lots of examples here. In fact, most 'basic' science is like this, the most famous example is surely Newton's law of gravity; Newton gave an expression which gave complete agreement with the motions of the planets around the sun, but with regards to what gravity actually is, he famously said *hypothesis non fingo*, which *very* loosely translated means 'don't ask me'.

(2) This phenomenon is almost entirely due to the Pauli principle; if the outer electron were to be described by a $1s$ AO, as in hydrogen, this AO would overlap strongly with the inner-shell $1s$ of the lithium atom, and violate the Pauli principle, as explained in Section 4.1.

Chapter 5

(1) Both the MO and the VB method have their strengths and weaknesses. Using HAOs the two models can be directly compared. It turns out that the VB model is more general and can give greater accuracy than the MO model, but at the expense of greater computational complexity.

For larger molecules, the VB method is increasingly impractical. Try to make sure that you are not tripped up by the two uses of 'hybrid'.

(2) The problem here is that one is trying to explain the behaviour of one molecule in the presence of another, and this means that the molecule's electron distribution *will* be distorted in the reaction. In the MO model this will change the form of the MOs, while in the VB model it will change the 'mix' of structures, so perhaps an ionic term will replace the covalent term as the dominant structure.

It is, perhaps, worth while to give a more detailed explanation of the comparison of the two models. The simplest and most direct way is to see how they compare for the simplest polar covalent bond, the Li–H single bond. We can use a single HAO on each atom.

Take two HAOs, one centred on each nucleus; say ϕ_1 on nucleus 1 and ϕ_2 on nucleus 2. They may be simply the s AO on each atom or, better, a polarised HAO — a linear combination of the s AO and a $2p_\sigma$ AO giving each HAO a better concentration of density between the nuclei.

The MO Model: The required MO (ψ, say) must have contributions from each HAO:

$$\psi = c_1\phi_1 + c_2\phi_2, \tag{1}$$

where c_1 and c_2 are the numerical co-efficients which give the optimum MO ψ.

The two-electron function for the whole molecule ($\Psi_{MO}(\vec{r}_1, \vec{r}_2)$, say) is then simply two electrons in this MO:

$$\Psi_{MO}(\vec{r}_1, \vec{r}_2) = \psi(\vec{r}_1)\psi(\vec{r}_2). \tag{2}$$

The VB Model: Using the same two HAOs, the VB model uses the three possible structures described in Section 5.5.1 on page 99 to form Ψ_{VB}, where:

$$\Psi_{HL} = \phi_1(\vec{r}_1)\phi_2(\vec{r}_2) + \phi_1(\vec{r}_2)\phi_2(\vec{r}_1) \tag{3}$$

$$\Psi_{+-} = \phi_2(\vec{r}_1)\phi_2(\vec{r}_2) \tag{4}$$

$$\Psi_{-+} = \phi_1(\vec{r}_1)\phi_1(\vec{r}_2) \tag{5}$$

These three functions represent, respectively, a covalent bond (LiH–H), one ionic 'bond' (LiH$^+$ H$^-$) and the other ionic bond (H$^-$ LiH$^+$), and the VB two-electron function is:

$$\Psi_{VB} = C_{HL}\Psi_{HL} + C_{+-}\Psi_{+-} + C_{-+}\Psi_{-+}. \tag{6}$$

The Comparison: The problem of comparing these functions is that Ψ_{MO} is written in terms of the MO (ψ) while Ψ_{VB} is written in terms of the HAOs (ϕ_1 and ϕ_2). But equation (1) gives us the relationship between these functions so we can use it to get *both* Ψ_{MO}) and Ψ_{VB} in terms of the same functions, and compare them directly.[1]

$$
\begin{aligned}
\Psi_{MO}(\vec{r}_1, \vec{r}_2) &= \psi(\vec{r}_1)\psi(\vec{r}_2) \\
&= [(c_1\phi_1(\vec{r}_1) + c_2\phi_2(\vec{r}_1)] \\
&\quad \times [(\phi_1(\vec{r}_2) + \phi_2(\vec{r}_2)] \\
&= c_1^2\phi_1(\vec{r}_1)\phi_1(\vec{r}_2) + c_2^2\phi_2(\vec{r}_1)\phi_2(\vec{r}_2) \\
&\quad + c_1c_2[\phi_1(\vec{r}_1)\phi_2(\vec{r}_2) + \phi_1(\vec{r}_2)\phi_2(\vec{r}_1)]
\end{aligned}
$$

But these are just the terms in the full expression for Ψ_{VB}:

(1) The term in square brackets multiplied by c_1^2 is Ψ_{-+}
    i.e. $C_{-+} = c_1^2$ in equation (6).
(2) The term in square brackets multiplied by c_2^2 is Ψ_{+-}
    i.e. $C_{+-} = c_2^2$ in equation (6).
(3) The term in square brackets multiplied by c_1c_2 is Ψ_{-+}
    i.e. $C_{HL} = c_1c_2$ in equation (6).

The conclusion of these manipulations is that Ψ_{MO} is of just the same form as Ψ_{VB} but *for a particular choice of the coefficients* C_{HL}, C_{+-} and C_{-+}. The question which remains is 'is this choice the best possible one for the function Ψ_{VB}?' The answer to this can be seen by the fact that, for the MO choice of these co-efficients,

$$C_{HL} = \sqrt{C_{+-}C_{-+}}.$$

That is, the three co-efficients are not *independent*. So, we conclude that the VB model is always in principle capable of giving a better answer than the MO model *using the same set of HAOs*. As we have noted, in practice the VB method is much more difficult to implement than the VB.

---

[1] You may wish to take this algebra on trust!

Chapter 6

(1) This is really a problem of symmetry; since the charges are equal they must arrange themselves as symmetrically as possible over the surface of a sphere.

(2) This is a case of trial and error but the rules should give you the right answer if you do the book-keeping right!

(3) Did you get it right? Look up the shape in Mark Winter's *Chemical Bonding* (Oxford University Press, 1994), which has more examples.

Chapter 7

First of all, as we shall see in detail in Section 14.3, simply 'counting' the electrons in a particular HAO within a molecule is a very coarse approximation to the electron distribution. All orbitals are functions of *space* and, just because an HAO is composed of AOs centred on a particular atom it does not necessarily mean that the electrons which it describes are close to that atom.

(1) Even a casual look at the contours of the σ and π bond MOs shows that the electrons in the σ bond lie mainly in the internuclear area while those of the π bond are:

    (a) Further from the internuclear area;

    (b) Shielded from the nuclei by the electrons in the σ bond,

and it is the attraction of the two nuclei which dominate the binding energy of the electrons. Notice that the fact that the σ electrons shield the ones in the π bond means that the π bond is less polar than the σ bond.

(2) Not really, all that one can tentatively conclude is the very obvious fact that the C=O bond is polar in the way that the relative electronegativities of the C and O atoms would suggest.

It must always be borne in mind that the calculations being described are for individual, isolated molecules at absolute zero of temperature. Real molecules in the laboratory are always vibrating and rotating which, of course, affects their measured dipole moment. If two molecules of the same chemical compound have different isotopes of the same element in their makeup, their vibrations and rotations, which depend on mass, will be different, and so should their measured properties. One might expect to find this effect most notable in molecules which have the mass of hydrogen atoms *doubled* by substitution of a proton for deuterium nucleus.

Chapter 9

The removal of an electron from any molecule (ionisation) is a large perturbation to the whole molecule. Even if the electron removed is not from a particular one of the substructures, this substructure and all others *will* be affected by the electrons re-organising themselves to reflect the new environment. The answers to the questions refer to what one might call the 'instantaneous' ionisation, the ideal situation where one removes an electron from one MO before the remaining electrons 'realise what has happened'.

The ionised electron can be represented as occupying an orbital *much* higher than any of the MOs in the diagram on page 166.

(1) There are only three distinct ionisations from the valence MOs: the σ-bond MO, a lone pair on a nitrogen atom and a π-bond MO.

(2) Only the removal of an electron from the σ-bond MO leaves a total electronic structure which has the symmetry of the homonuclear diatomic N_2. Removal of an electron from one of the two π MOs leaves the ion with the symmetry of interchanging the two N atoms, but it destroys the symmetry of the *two* π-bond MOs; one has only one electron and the other has two.

(3) The lowest ionisation energy is the energy required to remove an electron from the least tightly-bound MO, which in this case is the π-bond MO.

(4) The electrons in the two atomic cores in the N_2 molecule are the ones which are very tightly bound by the strong attraction of their own nuclei. They are the ones which will be least affected by the other electrons, and so there is the possibility that the removal of one of these electrons will generate an N_2 ion with a singly-occupied core AO which 'lives' long enough to be detected before an electron 'falls down' from one of the valence MOs to occupy the lower-energy core AO.

Chapter 14

We should not be disparaging about historical rules for helping with the understanding of molecular electronic structure. It is useful to remember that the theoretical tools which are really necessary to understand chemical bonding were not available to the scientists who struggled to formulate some kind of order among the structures of a huge collection of molecules. Any branch of science must start with attempts to systematise the data available, and the work of the early pioneers is absolutely crucial to further progress.

The octet rule was used with a great deal of success, particularly in organic chemistry, where the bonding patterns are more amenable to organisation and systematisation. The 18-electron rule played a similar rôle in bringing some order to the proliferation of transition-metal compounds, in which the metals have much more baffling multiple valencies.

The quantum theory of the hydrogen atom — the energies and AOs, together with the all-important Pauli principle, has deepened our understanding of molecular electronic structure. But a concentration on quantitative calculations sometimes involves the neglect of the fact that the bonding in molecules still involves the underlying rule that the complex electron densities which are revealed are still largely composed of relatively stable substructures: covalent and dative bonds, delocalised structures, atomic cores and lone pairs. We must beware of reacting too much against past explanations!

Chapter 16

There is another possibility for the structure: one of the most common reactions of a C=C double bond is where the π bond is broken and each carbon atom forms a new σ bond, thus producing a di-substituted saturated compound. Is it possible that a variant of this reaction is happening here, except that both the new σ bonds are to the same atom: the central platinum? This would, of course, form a molecule containing a three-membered ring. Such molecules are notoriously unstable when the ring is composed of three carbon atoms.

Index